THE ADAPTIVE ORGANIZATION

LEADING CHANGE IN THE AI ERA

Patrick McGarry

DATA LITERACY
PRESS

First edition published 2026
by Data Literacy Press
500 108th Ave NE, Suite 1100, Bellevue, WA 98004
https://dataliteracy.com

Library of Congress Control Number: 2026942237

Description: First edition. | Bellevue: Data Literacy Press, 2026

ISBN: 978-1-960907-12-7 (hardback)
ISBN: 978-1-960907-13-4 (ebook)

Cover design and interior graphics designed by Alli Torban

Printed in the United States of America.

MORE BY DATA LITERACY PRESS

Leading in the Age of Data
by Ben Jones, 2023

AI Literacy Fundamentals
by Ben Jones, 2024

Data Literacy Fundamentals, 2nd Edition
by Ben Jones, 2025

Chart Spark
by Alli Torban, 2023

For more information about this book,
including resources and updates, visit
https://mcgarrycdo.com/

CONTENTS

FOREWORD

by Douglas B. Laney

A chief data officer I advise recently told me about a moment of corporate consternation. Her team had spent eighteen months building what most practitioners would recognize as a solid data governance program including a catalog and classification schema, stewardship roles, access controls tied to sensitivity classifications, lineage documentation sufficient to satisfy audits. By my own or any other maturity assessment, her organization had leapt ahead of its peers. Then her budget review arrived and the CFO asked a question that ruptured the raison d'etre: *What are these data assets actually worth to us?*

This is a question I've been prodding executives to ask for the past couple decades. But she had no defensible answer, not because the work lacked value, but because no one had connected the governance architecture she had built to the type of economic calculus that encapsulates every other asset allocation decision throughout the company. The governance program had survived on the strength of a compliance argument alone. It wasn't until we started putting dollar signs on data—the replacement cost, revenue contribution, and other economic metrics for key data assets that business unit leaders began requesting "ownership" of them rather than resisting any responsibility or accountability for them.

I open with this story because Patrick McGarry's book addresses, with unusual depth and practitioner candor, the organizational architecture that this CDO and others have built. *The Adaptive Organization* is a practitioner's guide to designing organizations that can actually absorb and apply data, analytics, and AI rather than merely acquire them.

In particular, his CATALOG framework that you will learn about here treats culture, analytics, technology, alignment, leadership, operations, and measurement as interdependent disciplines rather than distinct workstreams managed by disparate committees. McGarry's insistence that *governance is infrastructure, not overhead*, is top-shelf thought that eludes most books on data strategy. And his application of Deming's quality economics to prove the arithmetic shows profound respect for other brilliant minds, especially that the cost of embedding governance from the start is always a fraction of the cost of retrofitting it after failures reach production.

For two decades, my own work on infonomics—the discipline of managing and measuring data as an actual economic asset—has argued that organizations cannot credibly call data "strategic" while forgoing its formal valuation or failing to apply the same asset management rigor they bring to physical and financial assets. The organizational structures McGarry describes make that logic operational. A well-designed CDO function does not merely prevent misuse; it creates the institutional capacity to extract and compound value from information assets over time. Without the organizational architecture this book details, data valuation remains an abstraction. With it, valuation becomes an operational discipline.

But the trajectory doesn't stop at data governance and valuation. This is just the foundational stuff of today's data-savvy organization. Rather, businesses that master what McGarry prescribes will be positioned to build what I have been calling the "self-driving organization," an enterprise that senses, decides, and acts with progressively greater levels of autonomy—from assisted recommendation, through automated execution within defined boundaries, to autonomous operations where swarms of AI agents set intermediate goals and select methods on their own. The most advanced frontier involves self-optimizing systems where multiple agents coordinate across functions, recalibrating resources and realigning strategy against enterprise-wide objectives.

None of that is possible without the foundation McGarry builds in these pages. Autonomous systems that lack semantic coherence, clear decision rights, and calibrated oversight will produce the kind of conspicuous failures that set entire programs back by years—or worse result in corporate collapse altogether. The governance-first architecture this book advocates is not merely compatible with the autonomous organization; it is proto-infrastructure for it.

Data, analytics, and AI professionals who read this book and build accordingly will have the infrastructure to reach these requisite levels of enterprise autonomy, where the compounding advantages become essential for business survival.

Patrick McGarry has produced a book that respects the reader's intelligence and the difficulty of the work. He doesn't promise transformation through technology acquisition, nor does he reduce leadership to platitudes. However, for anyone who suspects that the gap between data ambition and data reality is at its core an organizational design problem, this is the book that shows you how to close it. What you build from this guidance will determine not only whether your organization can govern data responsibly today, but whether it can operate tomorrow with the requisite autonomous speed, scale, and success.

— Douglas B. Laney, Data, Analytics and AI Strategist, and Author of *Infonomics and Data Juice*

PREFACE

I have spent my career at the intersection of people, technology, and data. From building open source programs and nurturing communities to designing enterprise systems and advising leaders on strategy, I have learned that data is never just about the bits and bytes. It is about the people and missions it serves. The work is both deeply technical and profoundly human.

Today, we (like so many generations before us) stand at an inflection point. The pace of innovation in data and AI is unlike anything we have seen before. Models are becoming more capable, platforms are becoming more integrated, and expectations are becoming more urgent. In only a few years, what once seemed impossible, such as linking fragmented datasets, automating governance, and operationalizing trust at scale, has become not only possible but expected. Yet many organizations, especially in the public sector, still struggle under the weight of legacy systems, outdated processes, and silos that were never designed for this new era.

I have had the privilege of working side by side with agencies and partners, most recently at data.world and ServiceNow, to help bridge this gap. Together, we have built architectures that not only manage data but also actively make it more useful, more trustworthy, and more aligned with mission outcomes. We have shown that modernization is not about replacing everything at once. It is about weaving together what already exists, applying the best of what is new, and creating the conditions where innovation can thrive.

My experience has taught me that strategy and execution cannot be separated; they are inextricably linked. It is one thing to articulate a vision for data and AI. It is another thing to sit down with engineers, analysts, and program leads and help them design the pipelines, ontologies, policies, and workflows that make that vision a reality. I have done both. That is why I believe the future of data requires leaders who can move fluidly between the big picture and the technical detail. Leaders who can inspire confidence at the boardroom table and, if not create a schema, at least understand the complexities and impact of that process at a rudimentary level.

This book is about the art of the possible. It is about helping you see what is coming, not as a threat to be feared, but as an opportunity to be harnessed. The next decade will be defined by how well we modernize our relationship with data and AI. Will we treat them as abstract tools, or will we embrace them as strategic assets that shape how we govern, defend, heal, and serve?

I write not only as an observer of this moment, but as a participant in it. I have seen how the right data systems, programs, and partnerships can transform an organization's ability to act. I have watched communities rally around open source innovation that changed the trajectory of entire industries. I have experienced the frustration of slow progress and the exhilaration of a team finally breaking through.

I hope that, as you read, you will feel the same sense of urgency and possibility that drives me. Because the truth is, the story of data is still being written. Every one of us has a role to play in shaping it.

WELCOME TO THE JOURNEY.

INTRODUCTION

Organizations across every sector face a common challenge: how to transform ambition about data and artificial intelligence into operational reality. The gap between what leaders say they want and what their organizations can actually deliver has never been wider. Executives declare themselves "data-driven" while decisions continue to rely on intuition and incomplete information. Boards approve AI initiatives while the underlying data remains fragmented, ungoverned, and misunderstood. Transformation programs launch with fanfare but stall when they encounter the friction of legacy systems, siloed teams, and cultures resistant to change.

This book is written for the leaders responsible for closing that gap. Chief Data Officers, Chief Analytics Officers, technology executives, and mission operators now carry the weight of organizational transformation in an era when data and AI are no longer optional enhancements but fundamental to how work gets done. The stakes are high. Organizations that succeed will gain decisive advantages in speed, insight, and adaptability. Those that fail risk irrelevance.

But success requires more than technology. It demands new ways of thinking about leadership, governance, and organizational design. The leaders who thrive will be those who understand that this transformation is as much human as it is technical, that strategy and execution cannot be separated, and that the choices made today will compound into the capabilities their organizations possess tomorrow.

The Nature of the Challenge

Every generation experiences technological change and tends to view it as unprecedented. In 1970, Alvin Toffler published *Future Shock,*[1] warning that the pace of change itself had become psychologically disorienting. His concerns, centered on television and jet travel, now seem almost quaint. Yet his core insight remains valid: living through transformation always feels more dramatic than reading about it afterward.

What distinguishes the current moment is not simply the presence of change but its character. Three forces are converging: capabilities are improving at exponential rather than linear rates, deployment timelines have compressed from decades to quarters, and the decisions being automated carry consequences that directly affect human welfare and institutional trust. Together, these forces create conditions that demand new forms of leadership.

The challenge is further complicated by the fact that AI infrastructure is being built in real time, under scrutiny, with immediate consequences. Unlike in previous technological waves that unfolded over generations, organizations must now make consequential choices about architecture, governance, and resource allocation while the landscape continues to shift beneath them. There is no playbook to follow because it is being written as practitioners learn what works and what doesn't.

The Emerging Role of Data Leadership

The Chief Data Officer role emerged from necessity, not tradition. Organizations discovered that treating data as a byproduct of operations rather than a strategic asset led to fragmentation, quality problems, and missed opportunities. Someone needed to be accountable for data as a discipline, with authority that crossed traditional functional boundaries. The

1. Toffler, Alvin. Future Shock. Random House, 1970.

CDAO role, encompassing both data and AI, represents the evolution of this thinking.

Yet the role remains poorly understood, even within organizations that have created it. Is the CDAO a technologist who speaks the language of infrastructure and pipelines? A strategist who connects data initiatives to business outcomes? A diplomat who brokers agreements between skeptical stakeholders? A risk manager who ensures governance and compliance? The answer, uncomfortably, is all of these. The CDAO must operate at the intersection of technology, strategy, culture, and policy, translating between worlds that rarely speak the same language.

This book traces the evolution of data leadership from its origins in IT departments through its current position at the executive table. It examines what effective data and AI leadership looks like in practice, the capabilities required to succeed, and the organizational conditions that allow these leaders to deliver impact. The role is still being defined, which means there is both opportunity and risk in how it evolves.

A Widening Divide

A gap is opening between organizations that have successfully integrated data and AI into their operations and those that have not. This divide is not simply about access to technology or budgets. It reflects bigger differences in how organizations think about data, how they structure accountability, and whether they treat governance as infrastructure or as an afterthought.

Organizations on the wrong side of this divide share recognizable patterns. They procure AI tools without addressing the data quality issues that will undermine them. They publish governance policies that never reach the workflows where decisions are actually made. They celebrate pilots that generate excitement but never scale because the organizational infrastructure to support them does not exist. Most critically, they treat data and AI as technology problems to be solved by purchasing software, when the real challenges are cultural, processual, and structural.

The consequences of falling behind are not abstract. Organizations that cannot integrate data and AI effectively will find themselves operating with inferior information, slower decision cycles, and diminishing ability to attract talent or compete for resources. Over time, the gap becomes self-reinforcing. Success compounds for those who get the fundamentals right, while organizations that struggle find themselves perpetually reacting rather than leading.

Closing this gap requires understanding what separates successful organizations from those that struggle. The differences are identifiable, which means they are also addressable. But they require deliberate effort and leadership commitment, not wishful thinking or superficial remedies.

Structure and Approach

This book is organized to address the full lifecycle of data and AI leadership, from understanding the strategic landscape through building capabilities and deploying them at scale. Each chapter is designed to be both conceptually grounded and practically useful, offering frameworks that leaders can adapt to their own contexts.

The opening chapters establish context. They examine why this moment feels different from previous technological shifts, what distinguishes organizations that succeed from those that struggle, and how the CDAO role has evolved to meet the demands of a data-centric world. These chapters provide the conceptual foundation for everything that follows.

The middle chapters focus on building capabilities. They explore how to construct AI strategies that balance ambition with realism, embed governance into workflows rather than treating it as compliance theater, cultivate innovation ecosystems that connect mission operators with technologists, and navigate the organizational politics that can derail even well-designed initiatives. The emphasis throughout is on deployable knowledge: playbooks, assessment frameworks, and decision models that translate concepts into action.

The later chapters address execution at scale. They examine deployment strategies that go beyond pilots, the architectural decisions that enable or constrain organizational agility, and the leadership practices that sustain momentum amid the inevitable setbacks and resistance that accompany transformation. These chapters draw on real examples from government, defense, healthcare, and enterprise contexts to illustrate what works, what fails, and why.

Throughout, the book confronts uncomfortable truths. Governance is often treated as an afterthought, leading to costly retrofits when problems surface. Organizations underinvest in their people, assuming that skills will develop on their own without deliberate development. Pilots fail not because the technology is inadequate but because the organizational readiness is absent. And perhaps most fundamentally, the economic logic of quality applies to data governance just as it did to manufacturing: the cost of building quality in from the start is always less than the cost of fixing failures downstream.

The Central Thesis

The argument that runs through this book can be stated simply: governance embedded in workflows from the outset is not overhead to be minimized but infrastructure to be invested in. This claim draws on principles articulated by W. Edwards Deming in the context of manufacturing quality.[2] Deming demonstrated that building quality into processes costs far less than inspecting for defects afterward, and catastrophically less than fixing failures that reach customers.

The same logic applies to data and AI governance. Organizations face a choice: invest in prevention by embedding governance into how work is designed and executed, or pay exponentially more later through compliance retrofitting, reputation damage, and mission failures. The arithmetic is unforgiving. Prevention costs are a fraction of detection costs, which are themselves a fraction of failure costs.

This is not an argument for bureaucracy or excessive control. It is an argument for intentional design. Governance that is embedded becomes invisible, enabling rather than constraining innovation. It creates the conditions where trust can compound, where experiments can proceed with confidence, and where organizations can move at the speed their missions demand. By contrast, governance treated as an afterthought becomes friction, slowing everything down while providing incomplete protection.

Organizations that internalize this logic join the ranks of those succeeding with data and AI. Those that treat governance as compliance

2. Deming, W. Edwards. Out of the Crisis. MIT Press, 1982.

theater, something to satisfy auditors but not integrate into operations, find themselves perpetually fighting fires and wondering why their initiatives stall.

Intended Audience

This book is written for practitioners, not theorists. It assumes the reader is responsible for delivering results in an organizational context where resources are constrained, politics are real, and failure has consequences. The primary audience includes Chief Data Officers, Chief Analytics Officers, Chief AI Officers, and the teams that support them. But the content is equally relevant for technology executives, mission operators, program managers, and anyone tasked with making data and AI work in practice.

The book is not a technical manual. It does not teach readers how to train machine learning models or architect data pipelines. Those skills are essential, but they are well covered elsewhere. Instead, this book addresses the harder problems: how to build organizational readiness, navigate skepticism and resistance, sequence initiatives for maximum impact, and sustain momentum when early wins fade, and the work becomes grinding.

Readers from government, defense, healthcare, financial services, and commercial enterprises will find relevant examples and frameworks. While contexts differ, the underlying challenges are remarkably consistent. Every organization struggles with legacy systems, siloed data, competing priorities, and the gap between strategic intent and operational reality. The specifics vary, but the patterns repeat.

How to Engage This Material

This book can be read sequentially or consulted as a reference, depending on immediate needs. The early chapters provide essential context and are best read in order for readers new to data and AI leadership. The middle chapters on strategy, governance, and innovation ecosystems can be read independently once the foundational concepts are understood. The later chapters on deployment and architecture assume familiarity with the frameworks introduced earlier.

Throughout the book, sidebars provide additional context, case examples, and practical tools. Playbooks offer step-by-step guidance for specific tasks such as conducting readiness assessments or structuring innovation competitions. These elements are designed to be extracted and adapted for use in real organizational settings.

The book deliberately avoids prescriptive templates. Every organization faces unique constraints, cultures, and missions. What works in a federal agency may fail in a commercial enterprise. What succeeds in healthcare may be inappropriate for defense. The goal is not to provide universal answers but to equip readers with the concepts, frameworks, and examples they need to make informed decisions in their own contexts.

Readers should expect to disagree with some arguments and to find gaps where their specific challenges are not addressed. That is intentional. The field of data and AI leadership is still evolving, and no single book can capture every nuance or anticipate every situation. The aim is to advance the conversation and provide scaffolding for leaders as they build their own approaches.

The Path Ahead

The transformation organizations face is neither optional nor distant. It is happening now, and the choices leaders make today will determine whether their organizations thrive or struggle. The good news is that the path forward is becoming clearer as more practitioners learn what works. The patterns of success are identifiable. The mistakes are predictable. The frameworks exist.

What remains is execution. That work is hard, often frustrating, and rarely linear. It requires patience when stakeholders resist, persistence when pilots fail, and the ability to maintain strategic clarity while navigating tactical chaos. But it is achievable. Organizations led by people who understand both the promise and the difficulty of this moment are succeeding. Their lessons form the foundation of what follows.

The chapters ahead offer a roadmap, but they do not guarantee success. That depends on leadership, organizational commitment, and the willingness to do work that is often unglamorous but always essential. The transformation is underway. The question is not whether to participate but how to do so effectively.

WELCOME TO THE WORK.

Chapter 1

THE DATA-DRIVEN TURNING POINT

It is not an exaggeration to say that data has become the new infrastructure of modern life. It runs beneath our feet like electricity, mostly invisible until the lights flicker. But this comparison deserves scrutiny, because it invites a critical question: Is AI infrastructure truly like electrical infrastructure, or is it something more fragile, more resource-intensive, and more contested?

The comparison to electricity implies permanence, ubiquity, and efficiency. Electrical grids, once built, become reliable foundations that enable everything else. Yet critics rightly point out that modern AI infrastructure, particularly the massive GPU farms required for training and inference, bears little resemblance to the elegant efficiency of a power grid. These facilities consume extraordinary resources: energy measured in megawatts, water measured in millions of gallons for cooling,[3] and rare earth minerals for chip fabrication. A single AI-generated image can require as much water for cooling as it takes to fill several water bottles.[4] Training a large language model can consume as much electricity as a small town uses in a month.

3. Li, Pengfei, et al. "Making AI Less 'Thirsty': Uncovering and Addressing the Secret Water Footprint of AI Models." arXiv preprint arXiv:2304.03271, 2023.
4. Patterson, David, et al. "Carbon Emissions and Large Neural Network Training." arXiv preprint arXiv:2104.10350, 2021.

This is not trivial. The environmental footprint of AI infrastructure is real, substantial, and growing. Critics who argue that GPU farms represent a kind of technological excess, burning resources to produce outputs of uncertain value, are not engaging in hyperbole. They are raising legitimate questions about sustainability, resource allocation, and whether the benefits justify the costs. Some go further, arguing that if the current AI wave stalls or crashes, these facilities will become stranded assets: expensive infrastructure optimized for a capability that proved less transformative than predicted.

I believe this critique, while important, misses the deeper trajectory. The comparison to electricity is not about resource efficiency today; it is about functional permanence. Early electrical generation was also wildly inefficient, dangerous, and contested. Edison's first power plants were local, small-scale, and required constant human intervention. Critics questioned whether electricity would ever be more than a luxury for the wealthy. Yet once the capability existed, once people experienced what electricity enabled, there was no going back. The question became not whether to have electrical infrastructure, but how to make it more efficient, more reliable, and more accessible.

AI infrastructure is following a similar path, though the timeline is compressed. Yes, today's systems are resource-intensive. Yes, the cost-benefit calculus is still being worked out. Yes, there are legitimate questions about whether every use case justifies the environmental burden. But the capability itself, the ability to process language, recognize patterns, and generate insights at scale, has crossed a threshold. Organizations and individuals who have experienced what these capabilities enable are not willing to revert. The question is no longer whether AI becomes infrastructure, but how quickly we can make that infrastructure more efficient, more sustainable, and more aligned with societal values.

The environmental concerns are not background noise to be dismissed. They are urgent engineering and policy challenges that must be addressed. Moore's Law has historically delivered efficiency gains alongside capability gains; more computation per watt, year after year. The same pattern must hold for AI infrastructure, or the environmental costs will indeed become prohibitive. Research into more efficient architectures, better cooling systems, renewable energy integration, and selective deployment (using AI where the impact justifies the cost) is not optional. These advances are prerequisites for AI infrastructure to achieve the permanence that electricity achieved.

So when I describe data and AI as infrastructure, I am making a claim about functional necessity, not resource efficiency. The work ahead is to close the gap: to make AI infrastructure as reliable, efficient, and justifiable as the electrical grid it depends on. That work is underway but incomplete. Acknowledging this tension honestly is essential because organizations that treat AI as "infrastructure" without grappling with its current costs risk both environmental harm and reputational damage.

TREND SNAPSHOT
The OECD's AI Principles

In 2019, the Organisation for Economic Co-operation and Development (OECD) adopted five high-level principles for AI[5], later endorsed by the G20. These principles remain a cornerstone of international guidance:

1. AI should benefit people and the planet.
2. AI systems should be designed with respect for human rights and democratic values.
3. Transparency and explainability must be prioritized.
4. Robust safety and security protections are essential.
5. Accountability is non-negotiable.

Why this matters: These principles are echoed across policy frameworks worldwide, from the EU's AI Act to voluntary U.S. guidelines. For organizations in every sector, aligning with them is not just a compliance exercise; it is a signal of trustworthiness.

5. Organisation for Economic Co-operation and Development. OECD Principles on Artificial Intelligence. OECD, 22 May 2019. https://oecd.ai/en/ai-principles .

Contested Infrastructure and the Path Forward

Infrastructure is never neutral. Highways enabled suburbanization but also displaced urban communities. The electrical grid powered industrial growth, but also locked in carbon-intensive generation for decades. The internet democratized information but also enabled surveillance capitalism and misinformation. Every infrastructure choice creates winners and losers, enables some futures while foreclosing others, and embeds values, often invisibly, into the systems that shape daily life.

AI infrastructure is no different, except that it is being built in real time, under scrutiny, and with immediate consequences rather than delayed ones. This creates an opportunity that previous infrastructural moments lacked: the chance to shape the infrastructure intentionally rather than inheriting the unexamined assumptions of early builders.

The environmental costs are one dimension of this contestation. Others include: Who controls the infrastructure? If AI capabilities concentrate in the hands of a few hyperscale cloud providers, does that create dependencies that undermine organizational autonomy or national sovereignty? What happens to smaller organizations, research institutions, or nations that cannot afford access? How do we ensure that the infrastructure serves public interest alongside commercial interest?

These are not abstract questions. They shape whether AI becomes a force for broad-based opportunity or concentrated advantage, whether it reinforces existing inequalities or creates new pathways to capability, and whether it operates under democratic accountability or escapes it entirely. The answers will not come from technology alone. They require policy, governance, investment in alternatives (such as open-source models, public compute resources, and efficiency breakthroughs), and sustained attention to the distributional consequences of infrastructural choices.

Organizations building AI strategies today are not simply adopting a tool. They are participating in the construction of contested infrastructure. The choices they make (about where to source compute, how to measure environmental impact, whether to support open alternatives, and how transparently to operate) will compound into the infrastructure that their successors inherit. This is infrastructure with implications, and treating it as a settled fact rather than an ongoing construction project is a mistake. The work is to shape it deliberately, not to accept it passively.

The Gap Between Ambition and Reality

Despite real progress over the past decade, a sobering gap persists between the "haves" and the "have-nots". Nearly every organization now names data and AI as top priorities, yet many still struggle to convert ambition into impact. Declaring "we are data-driven" is easier than making it true in daily decisions. Data remains locked in silos, AI tools are procured without transparency, and governance lives as PDFs instead of embedded controls.

This gap is not apathy. It is misaligned incentives, incompatible stacks, and a widening skills divide. Above all, it is socio-technical. Technology can unlock possibilities, but without the right incentives, cultural norms, and governance, those possibilities stall. Where progress is real, leaders behave differently. Organizations that treat data and AI as both cultural and technical transformations embed governance, invest in literacy, and build resilient pipelines. Those who treat them as software purchases often end up with tools, or worse yet "toys", and no outcomes.

The Data Haves and the Data Have-Nots

The "haves" of data and AI modernization invest in culture, governance, and infrastructure, not just tools. They integrate across silos, trust their sources, and operationalize AI at scale. The "have-nots" may possess data, but it is fragmented, talent is scarce, and governance is reactive.

This contrast is stark in retail. When we examine Walmart, their real-time inventory systems predicted demand at the store level and proved decisive during pandemic-era supply shocks. Competitors without comparable capabilities faced stockouts and delays despite the same external conditions. Capability, not circumstance, explained the difference.

The pattern repeats in healthcare. Systems with interoperable platforms forecast surges and reallocated staff; those without access or integration struggled to respond at speed. The divide was not just efficiency. It affected patient outcomes.

Tactical Outcomes: What "Haves" Get Right

The haves share four strengths: They (1) integrate data horizontally and vertically to create a single view of operations, (2) embed governance in workflows, so compliance does not slow innovation, (3) scale experiments directly into enterprise capabilities, and (4) create a baseline of data literacy that equips decision-makers to use AI critically. By contrast, organizations that lag exhibit the inverse: Silos persist. Rules are applied inconsistently and late. Pilots never scale. Trust erodes, leading either to blind acceptance of AI outputs or blanket rejection.

The Economics of Embedded Governance

The second strength, "embedding governance in workflows", deserves particular attention because it challenges a deeply held misconception about the cost of quality. W. Edwards Deming, the quality management pioneer who transformed post-war Japanese manufacturing, argued that the cost of improving quality is less than the savings it yields.[6] This principle, which revolutionized how organizations think about manufacturing defects, applies equally to data and AI governance.

Organizations often view governance as expensive overhead: compliance officers to hire, approval processes to navigate, review boards to convene, and documentation to maintain. Leaders see friction that slows innovation. But this view miscalculates the true economics. Deming taught that building quality into processes from the start costs far less than inspecting for defects afterward, and catastrophically less than fixing failures that reach customers.

In data and AI, the cost structure follows the same pattern:

Prevention costs (building governance in): Standardized data-sharing agreements, clear evaluation criteria, security requirements designed into architecture, training for responsible AI use, and playbooks that capture institutional knowledge.

Detection costs (inspecting for problems): Mid-project compliance reviews, data quality audits, model validation testing, and security

6. Deming, W. Edwards. Out of the Crisis. MIT Press, 1982.

assessments of deployed systems. These typically cost 10 times as much as prevention costs.

Internal failure costs (fixing before deployment): Retrospective compliance retrofitting, data cleanup for contaminated models, rebuilding systems that lack audit trails, and rework when pilots can't scale due to governance gaps. These typically cost 100 times as much as prevention costs.

External failure costs (fixing after deployment): Data breaches that destroy trust, biased algorithms that create legal liability, unauthorized data use that triggers regulatory penalties, and mission failures when operators can't trust AI outputs. These costs are often catastrophic and irreversible.

The "haves" understand this arithmetic intuitively. They invest in prevention because they have learned that every dollar spent on embedding governance into workflows saves ten dollars in compliance retrofitting and one hundred dollars in failure remediation. Compliance does not slow their innovation; it accelerates it by removing uncertainty about what is permissible and creating the conditions where trust can compound over time.

The "have-nots," by contrast, treat governance as inspection. They procure AI tools first and worry about compliance later. They launch pilots without establishing data provenance. They deploy models without documentation of training data or performance characteristics. When problems inevitably surface, they discover that retrofitting governance into live systems is both expensive and incomplete. Some problems, once embedded in production, cannot be fully remediated without starting over.

Deming's insight was that organizations cannot afford not to invest in quality. The same is true for data governance. The question is not whether you will pay for governance, but when you will. Pay upfront through thoughtful design, or pay exponentially more through firefighting, reputation damage, and lost opportunity. The arithmetic is unforgiving, and the choice determines whether an organization joins the haves or remains among the have-nots.

Implications for Individuals, Teams, and Organizations

These choices cascade down to people and teams. For individuals, the haves provide better tools and clearer information; the have-nots often resort to guesswork. For teams, a shared source of truth accelerates collaboration, while conflicting reports drain time and trust. For organizations, the haves pivot quickly and widen the gap over time. Research consistently shows that companies treating data as a strategic asset are far more likely to attribute significant earnings impact to analytics. As Tomasz Tunguz and Frank Bien note in "Winning With Data",[7] persistent "data brawls" inside organizations are a symptom of missing alignment, not missing dashboards.

For organizations as a whole, the divide determines resilience. "Data-haves" can pivot quickly, adapting to shocks and opportunities alike. Data "have-nots" are reactive, constantly playing catch-up. Over time, the competitive distance between these groups widens. As McKinsey has documented, companies that treat data as a strategic asset are three times more likely than others to say that their data and analytics contributed at least 20% to earnings.[8]

Pitfalls to Watch For

Avoiding regression requires vigilance. Common traps include buying advanced AI before addressing data quality and governance, publishing policies that never reach workflows, assuming skills will materialize without training, and celebrating pilots that never scale. Superficial "AI readiness" surveys compound the problem. The work is introspective. Technology, people, and processes must be aligned to use AI responsibly and effectively.

Each of these pitfalls reflects what Deming called "dependence on inspection to achieve quality", the costly practice of building systems first and checking them afterward. Buying advanced AI before addressing data quality is the equivalent of manufacturing defective products and then sorting the bad from the good. Publishing policies that never reach workflows is governance theater: documents that make leaders feel protected

7. Tunguz, Tomasz, and Frank Bien. Winning with Data: Transform Your Culture, Empower Your People, and Shape the Future. Wiley, 2016.
8. McKinsey & Company. Catch them if you can: How leaders in data and analytics have pulled ahead. McKinsey & Company, 19 Sept. 2019. McKinsey.com.

while providing no actual protection. Celebrating pilots that never scale is organizational waste: the innovation equivalent of scrap and rework.

The path forward requires a fundamental reframing. Governance is not overhead to be minimized; it is infrastructure to be invested in. Quality is not something you inspect for; it is something you design into your processes. The organizations that thrive in the AI era will be those that understand Deming's lesson: the cost of building quality and governance into your systems from the start is always less than the cost of fixing failures after they occur. This is not a theory. It is economics. And it is the difference between the haves and the have-nots.

CASE IN POINT
Applying Deming's Quality Economics to Data Governance

W. Edwards Deming taught that "quality comes not from inspection but from improvement of the process." His cost-of-quality framework divided expenses into four categories:

Prevention: Designing processes to avoid defects

Appraisal: Inspecting products to detect defects

Internal failure: Fixing defects before delivery

External failure: Fixing defects after delivery

Deming demonstrated that prevention costs are a fraction of appraisal costs (typically 10x less), and appraisal costs are a fraction of failure costs (typically 10x less again). The savings from reducing downstream costs far exceed the investment in prevention.

For data and AI governance:

Prevention = Governance embedded in workflows from day one

Appraisal = Audits and compliance reviews of existing systems

Internal failure = Retrofitting governance, remediating quality issues

External failure = Data breaches, biased algorithms, regulatory penalties

LESSON: Organizations cannot afford not to embed governance. The cost of prevention is always less than the cost of failure. The "haves" understand this arithmetic; the "have-nots" learn it the hard way.

The "AI-Ready" Imperative

"AI-ready" is not jargon. It reflects the reality that AI is a present-tense capability that requires preparation across technology, culture, processes, and governance. Readiness looks different by context. A government agency emphasizes secure, cataloged access to sensitive data. A consumer company prioritizes fairness and transparency to protect trust. A hospital focuses on system integration to protect patient safety. In every case, software alone is insufficient. The task is to create conditions that enable AI to operate responsibly across the enterprise.

Many organizations underestimate the depth of the challenge. Quick surveys and box-checking do not reveal whether people can use AI responsibly, whether processes absorb AI insights, or whether systems can meet governance demands. Concise, recurring assessments are helpful, provided they are grounded in the organization's culture and goals, rather than generic templates. Readiness shows up in three places that reinforce one another:

People. Data literacy anchors adoption. Without it, organizations oscillate between uncritical reliance and blanket rejection. Train staff to question outputs, understand limits, and apply human judgment.

Processes. AI must be placed at the decision point to matter. Redesign workflows so predictions arrive when action is possible, not after commitments are locked.

Systems. Readiness requires knowing where critical data resides, who is responsible for it, and how it is managed. Defense, retail, and nonprofit contexts differ, but the constant is a trustworthy pipeline. Deming's third point, "cease dependence on inspection to achieve quality", applies directly here. Organizations that inspect data quality after collection will always lag those that build quality into data-capture processes. The goal is not perfect data, which is impossible, but systems designed to maintain known quality levels and flag deviations automatically.

That said, even with the right focus, common pitfalls remain.

Pitfalls to Avoid

The biggest pitfall is treating readiness as a one-time certification. It is not. AI readiness is dynamic. As technologies evolve, as regulations shift, and as missions or business priorities change, organizations must continually reassess their position. A company that was AI-ready five years ago, before the explosion of generative models and new privacy laws, may not be AI-ready today.

Another pitfall is thinking that readiness can be outsourced entirely to consultants or integrators. Outside expertise can be helpful, but no external party can fully understand your organization's culture, incentives, and context. True readiness comes only from internal reflection and stakeholder engagement.

Finally, there is the trap of overemphasizing technology at the expense of people and processes. An organization can have the most advanced machine learning infrastructure in the world. Still, if its decision-makers don't trust it, or if the workflows don't accommodate it, it will remain unused.

All of this leads to a different way of thinking.

From Checklist to Conversation

What makes AI readiness so challenging, and so important, is that it is not a checklist; it is a conversation. A conversation between leadership and staff, between policy and technology, between ambition and reality. The organizations that succeed will be the ones that ask hard questions, adapt their processes, and engage their people in building a culture of responsible innovation.

In short, AI readiness is not about being ready to *buy* AI; it's about being prepared to *live with it*.

Why This Moment Feels Different

Context matters for why these choices feel urgent. Every generation experiences rapid change and tends to view it as unprecedented. In the 1990s, the rise of the commercial internet felt like a revolution. In the 2000s, the move to cloud computing promised to upend decades of enterprise IT practice. In the 2010s, the smartphone era transformed how individuals lived, worked, and connected. Each of these shifts reshaped the landscape of technology and society, and each carried its own mix of hype, opportunity, and disruption.

And yet, today's inflection point, centered on data and AI, feels qualitatively different. Not because it is bigger, faster, or noisier, though in some respects it is all of those things. It feels different because it represents a convergence of multiple forces: mature infrastructure, democratized tools, and evolving governance. Together, they create conditions where AI adoption is not just an option but an inevitability.

Lessons from the Dot-Com Era

The dot-com boom of the late 1990s offers a useful comparison. At that time, the internet was touted as a force that would eliminate friction, transform industries, and empower consumers. Some of those promises were overblown (remember the grocery delivery startups that collapsed under their own weight?), but many proved true. Amazon, eBay, and Google were born during that era, and the world has not looked the same since.

The similarities to today are striking: we see breathless headlines about AI's potential, massive venture capital flowing into startups, and organizations scrambling to avoid being "left behind." But there are also critical differences. The internet required extensive infrastructure build-out, including laying fiber-optic cables, creating server farms, and designing new protocols. AI, by contrast, rides on top of existing infrastructure. Thanks to the cloud, the hardware investment has (mostly, except for massive amounts of money spent on GPUs to NVIDIA's delight) already been made, and the tools are available at a fraction of the cost. That means adoption can scale far faster.

The pitfall then, as now, is mistaking early hype for guaranteed transformation. Just as many dot-com companies failed because they lacked viable business models, many AI pilots today will collapse under the weight of unrealistic expectations and business models that ultimately don't match the underlying costs. The lesson for organizations is not to reject the hype, but to separate durable value from temporary excitement.

A later wave offered a different lesson:

Modernization Waves and the Cloud

The early 2000s brought another tectonic shift: modernization and the rise of cloud computing. Enterprises that had long relied on sprawling, on-premise data centers began migrating workloads to the cloud. This was not just a technical migration; it was an organizational reckoning. CIOs had to rethink procurement, compliance officers had to reimagine security models, and finance departments had to adjust from capital expenditures to operational expenditures.

We're still feeling the repercussions of the security concerns around cloud, as the arms race hasn't slowed down yet. Public sector agencies attempt to address this through rigorous controls and programs like FedRAMP / StateRAMP, but the tension between convenience and security doesn't appear to be easing anytime soon. Given the continued rise of bad actors, the security landscape, and the brute force of social engineering attacks, it is a constantly evolving battlefield. We're likely to only see this arms race accelerate in operational tempo, given what AI can do for both sides.

Cloud adoption took years because it required this reconfiguration of culture and policy, not just systems. AI adoption today faces a similar challenge. The models may be ready to use, but organizations must rethink their workflows, retrain their people, and adapt their governance. The difference is speed. What took a decade with the cloud is happening in real time with AI, compressed into months or even weeks. That speed makes readiness and responsibility even more critical.

The Democratization Factor

Perhaps the most striking difference between now and previous waves is the degree of democratization. When the internet emerged, technical barriers were high. Building a website required coding skills and access to a server. In the early cloud era, moving workloads required teams of architects and long procurement cycles. By contrast, today's AI tools are available to anyone with an internet connection. A student can experiment with large language models in their dorm room. A small nonprofit can deploy image recognition tools without ever purchasing hardware.

This democratization is both exhilarating and dangerous. It lowers the barrier to experimentation, but it also raises the stakes for governance. Inexperienced users can deploy powerful models without understanding their limitations, leading to errors, bias, or even harm. The operational guidance here is clear: democratization must be matched by education and oversight. Just as the internet required the rise of digital literacy, AI requires the rise of AI literacy.

Policy as a Co-Evolving Force

Another critical difference in this moment is the role of policy. In past waves, regulation often lagged far behind innovation. The internet was largely unregulated in its early years, creating both extraordinary opportunities and significant risks, from online fraud to disinformation. The cloud, too, outpaced policy, forcing regulators to catch up with new questions around data sovereignty, privacy, and security.

With AI, policy is emerging alongside the technology. The European Union's AI Act, the U.S. National AI Research Resource pilot, and frameworks from the OECD and NIST are all attempts to set guardrails in real time. This co-evolution of technology and policy is unprecedented. For organizations, it means that compliance and governance cannot be afterthoughts; they must be part of the design from day one. As access expanded, oversight evolved in parallel.

Pitfalls of Misreading the Moment

There are two major pitfalls organizations face in misreading why this moment is different. The first is overconfidence, assuming that because AI tools are accessible, they are also easy to integrate responsibly. The second is fatalism, the belief that the changes are so overwhelming that nothing can be done to prepare. Both extremes are dangerous. The reality is that AI is powerful but not magical, transformative but not automatic. Organizations that succeed will be those that treat readiness as a journey, not a destination, and who balance experimentation with reflection.

Guidance for Leaders

For leaders navigating this moment, the parallels to past shifts offer useful guidance. From the dot-com boom: beware hype, but don't ignore transformative potential. From cloud modernization: recognize that culture and process change matter as much as technology. From democratization: embrace accessibility, but invest in literacy and governance. And from the policy wave: stay ahead of regulation, not behind it.

This is why today feels different. It is not just another technology cycle. It is the convergence of mature infrastructure, democratized tools, and evolving governance. Together, these forces make AI not simply a technology to adopt, but a context to live within.

TREND SNAPSHOT
Echoes of Past Technology Shifts

- **Dot-Com Boom (1990s):** High hype, low barriers to entry, but many failures. Durable winners focused on solving real problems at scale.
- **Cloud Modernization (2000s):** Long, complex migrations requiring cultural and financial shifts. Rewards flowed to those who rethought processes, not just systems.
- **Smartphone Era (2010s):** Democratization of access changed behavior faster than organizations could adapt. Apps reshaped expectations in every sector.
- **AI Today (2020s):** Convergence of mature infrastructure, democratized tools, and evolving governance. The speed of change is compressed, but so are the risks.

TAKEAWAY: *Each wave brought opportunity and disruption. AI is no different, but its convergence of forces makes the consequences broader and faster–moving than ever before.*

A Shared Responsibility

When people talk about AI, it is tempting to frame the conversation narrowly: either as a technical problem to be solved by engineers or as a policy problem to be addressed by regulators. But, as stated before, this deployment and utilization of associated data and systems is a socio-technical challenge, which is why the idea of shared responsibility is so important. Building AI systems that are trustworthy, ethical, and effective cannot be the job of a single team or department. Nor can it be outsourced entirely to vendors or consultants. It requires collaboration across people, processes, and technology. Each dimension matters, and each brings risks if neglected.

People: Culture and Accountability

At the heart of the socio-technical equation are people. AI is not self-governing; it reflects the values, incentives, and biases of those who train, build, and use it. This means that readiness is not just about technical skills but about cultivating a culture of accountability and shared ownership.

Consider a healthcare organization adopting predictive analytics to anticipate patient needs. If clinicians distrust the models, they may ignore the insights they provide. If administrators push adoption without addressing those concerns, they risk alienating staff and undermining patient safety. A responsible approach requires engagement across stakeholders (clinicians, data scientists, ethicists, and patients themselves) to ensure the system supports, rather than supplants, human judgment.

Culture only sticks, however, when it is translated into repeatable steps.

Process: Embedding Trust in Operations

Processes provide the scaffolding for responsible AI use. Without clear processes for data stewardship, for example, organizations risk deploying models on incomplete or biased datasets. Without oversight processes, unintended consequences may go unnoticed until they cause harm.

Operational guidance here is straightforward but not simple: build governance into the workflow, not as an afterthought. For instance, if an organization uses AI to screen job applicants, there should be a repeatable process for auditing the training data, testing the model for bias, and reviewing outcomes over time. This cannot be a one-time check; it must be institutionalized.

The consequences of failing to embed governance early are not theoretical. Cathy O'Neil's "Weapons of Math Destruction" documented how algorithms deployed without adequate oversight perpetuate and amplify existing inequalities at scale.[9] O'Neil examined criminal justice risk assessment tools that encoded racial bias, teacher evaluation systems that punished educators based on flawed statistical models, and credit scoring algorithms that trapped people in cycles of high-interest debt. In each case, the models were deployed first and questioned later, if at all. By the time problems surfaced, thousands or millions of people had already been affected.

What made these systems particularly destructive, O'Neil argued, was their opacity combined with their scale. The algorithms were opaque: people affected by them could not see how decisions were made or challenge their logic. They operated at scale: a single flawed model could affect millions of loan applications, job screenings, or parole decisions. And critically, they lacked feedback loops: the people designing the models rarely saw the human consequences of their errors, so the systems were never corrected.

O'Neil's work demonstrates why governance cannot be retrofitted. Organizations that deploy AI screening tools for hiring, creditworthiness, or healthcare triage without building in bias testing, explainability, and feedback mechanisms are not just taking technical risks. They are creating what O'Neil calls "mathematical weapons of destruction" that can systematically disadvantage vulnerable populations while appearing objective because they are backed by data. The damage is both individual (people denied opportunities or services unfairly) and institutional (loss of public trust, regulatory intervention, reputational collapse).

The operational imperative is clear: governance must be designed into AI systems from the beginning, not audited after deployment. This means establishing processes for examining training data for bias, testing model outputs against demographic groups, building in explainability so decisions can be challenged, and creating feedback loops that capture

9. O'Neil, Cathy. Weapons of Math Destruction: How Big Data Increases Inequality and Threatens Democracy. Crown, 2016.

real-world outcomes. When governance is embedded, problems surface early when they can still be fixed. When governance is an afterthought, organizations discover their mistakes only after causing widespread harm.

Technology: Open and Interoperable Systems

On the technology side, openness and interoperability are essential. Closed systems may offer short-term efficiency, but they create long-term risk: vendor lock-in, brittle integrations, and difficulty scaling. Open systems, by contrast, encourage collaboration and innovation by allowing organizations to build on one another's work.

Public health offers a compelling example. During the COVID-19 pandemic, agencies that relied on interoperable data standards were able to quickly share case counts, lab results, and vaccination data across jurisdictions. Agencies without such systems struggled, leading to delays and inconsistencies in reporting. The difference was not just technical. It was a matter of public trust and effective response.

Collaboration Across Stakeholders

Shared responsibility also extends beyond the walls of any single organization. The most pressing challenges—climate change, national security, healthcare, and economic inequality—cannot be solved by one entity acting alone. Collaboration across sectors is not a luxury; it is a necessity.

For example, in manufacturing, AI-powered predictive maintenance often requires collaboration among equipment makers, operators, and service providers. Each holds part of the data puzzle. Only by sharing data in agreed-upon formats and under clear governance can the full benefits be realized. Similarly, civic technologists building tools for government transparency rely on open data from public agencies. When that data is locked in proprietary systems or released in inaccessible formats, the entire ecosystem suffers.

Pitfalls of a Narrow Approach

The most common pitfall is treating AI as a purely technical solution. Deploy a model, write some code, automate a workflow, problem solved. But without parallel investments in people and processes, these solutions rarely stick.

Another pitfall is allowing systems to remain closed, either by design or inertia. Closed systems may serve narrow organizational goals, but they limit the broader ecosystem's ability to innovate and respond. Finally, there is the risk of fragmentation: every team or agency pursuing its own siloed approach, leading to duplication of effort and conflicting results.

From Responsibility to Collective Action

The real opportunity lies in moving from responsibility as a burden to responsibility as a collective act. Shared responsibility does not mean diluting accountability. It means recognizing that accountability is distributed, each actor has a role to play, and the system works best when those roles are aligned.

When public health agencies break down silos to respond to outbreaks faster, or when manufacturers reduce waste through predictive maintenance, or when civic technologists create open-source tools for transparency, they are not isolated examples. They are proof points that collaboration across people, processes, and technology can deliver outcomes greater than the sum of their parts.

The deceptively simple question that emerges from these examples (how do we turn the promise of AI and data into practice, ethically, sustainably, and at scale) can only be answered through shared responsibility. The path forward is not about building the perfect tool or writing the perfect regulation. It is about building resilient, open, and inclusive ecosystems.

CASE IN POINT
Apple: Control Meets Collaboration

Apple is often held up as the archetype of a closed, tightly controlled technology ecosystem. From the Macintosh to the iPhone, the company built its reputation on end-to-end integration of hardware, software, and services. This tight control allowed Apple to deliver elegant design and seamless user experiences that competitors struggled to replicate. At first glance, it might appear that Apple succeeded precisely because it "did everything itself."

But the reality is more nuanced, reflecting a tension Eric Raymond explored in "The Cathedral and the Bazaar." Raymond contrasted two development models: the cathedral, where software is carefully crafted by a small group of experts working in isolation, with releases at long intervals, and the bazaar, where development happens in public, with contributions from anyone who wants to participate.[10] Raymond argued that the bazaar model, exemplified by Linux, often produces better software because "given enough eyeballs, all bugs are shallow." More contributors mean more people finding and fixing problems.

Apple's success demonstrates that the choice between cathedral and bazaar is not binary. Apple's success has always relied on collaboration and openness, just in carefully chosen ways. Its operating systems are built on Darwin, a derivative of the open-source BSD Unix. Safari's WebKit engine was open-sourced and later became the basis for Google Chrome. Apple has contributed to projects such as LLVM and has released its Swift programming language as open source. The iPhone itself became indispensable not only because of Apple's design, but also because of the millions of third-party developers who built apps within its curated ecosystem.

Apple operates as a cathedral for core design decisions (hardware integration, user experience, platform architecture), while running a bazaar for extending functionality (third-party apps, open-source components, developer contributions). This hybrid approach captures Raymond's insight that "good programmers know what to write, great ones know what to rewrite (and reuse)." Apple reuses open-source components when they add no competitive differentiation, allowing the company to focus its resources on areas where design and integration create unique value.

The lesson for data and AI leaders is clear: even the most design-driven,

10. Raymond, Eric S. The Cathedral and the Bazaar: Musings on Linux and Open Source by an Accidental Revolutionary. O'Reilly, 1999.

controlled companies do not succeed in isolation. Apple's distinctiveness comes not from avoiding collaboration, but from strategically blending it; leveraging open standards and communities where they create scale (the bazaar), and differentiating through design and curation where it matters most (the cathedral). Organizations building data and AI capabilities must make similar choices: which components benefit from broad collaboration and which require tight control.

An Invitation to Collaborate

As emphasized in the introduction, the future of data and AI will not be built in isolation. It will be built in the open, by people and organizations willing to share what works, admit what does not, and learn together in real time. This spirit of collaboration is not just a nice-to-have. It is the only way to solve the complex, socio-technical challenges that define this moment.

Think back to the most significant advances in technology and science over the past century. The Human Genome Project succeeded not because one lab tried to sequence the genome in secret, but because governments, universities, and private companies pooled their efforts, agreed on data standards, and made results publicly available. The development of the World Wide Web itself was built on openness: common protocols, open standards, and the willingness of thousands of people to build and share tools. In contrast, technologies that have remained siloed and proprietary often stagnate or become brittle over time, unable to scale beyond narrow use cases.

Data and AI fall firmly into the first category. Their potential is too large, and their risks too complex, for any one organization to handle alone. Collaboration must extend across organizational boundaries, disciplines, and even national borders.

Collaboration Across Boundaries

Collaboration begins with recognizing that no single organization holds all the necessary data, talent, or insight. For example, public health agencies may have surveillance data, hospitals may have patient records, researchers may bring modeling expertise, and private companies may contribute technology platforms. Only by bringing these elements together can societies mount effective responses to pandemics or chronic health crises.

In the private sector, collaboration may take the form of competitors joining forces to set interoperability standards that benefit the entire industry. Think of how the automotive sector collaborated on standards for electric vehicle charging (even when it was painful at times, it still yielded mostly beneficial outcomes and interoperability). Each company benefits individually, but the real benefit comes from growing the ecosystem together. AI requires a similar mindset: agreement on standards for explainability, ethics, and governance will ultimately strengthen, not weaken, competitive advantage.

What Collaboration Looks Like in Practice

For individuals, collaboration means cultivating openness to new ideas and being willing to share lessons learned, even failures. A data scientist who openly documents model limitations helps not just their team, but anyone else working on similar problems.

For teams, collaboration means designing processes that assume data and knowledge will be shared rather than hoarded. This could mean adopting open APIs, using interoperable data formats, or building joint dashboards that provide a shared source of truth.

For organizations, collaboration means participating in broader ecosystems, contributing to industry standards, engaging in multi-stakeholder forums, and making portions of data open when appropriate. The open data movement has shown that transparency can fuel innovation far beyond what any one actor could achieve alone.

Pitfalls in Collaboration

Of course, collaboration is not without its challenges. One pitfall is performative openness, talking about collaboration while keeping systems locked down and data inaccessible. Another is uneven participation, where some stakeholders contribute heavily while others "free ride" on the group's efforts. Yet another is the absence of trust, where stakeholders suspect each other's motives and therefore hold back.

These pitfalls are real, but they are not insurmountable. The key is intentional design: establishing governance structures that fairly distribute responsibilities, building technical systems that enable safe sharing, and investing in trust-building among partners.

Collaboration as a Mindset

Ultimately, collaboration is not just a tactic; it is a mindset. It requires humility, the recognition that no one has all the answers. It requires curiosity, the willingness to learn from other fields and perspectives. And it requires courage, the bravery to admit missteps publicly so that others can avoid them.

The chapters that follow are written in this spirit. They do not offer prescriptive formulas that can be copied and pasted. Instead, they provide frameworks, lessons, and approaches meant to be adapted, remixed, and improved in your own context. They are contributions to a conversation that must remain open, iterative, and collective.

Success, in this sense, will not be measured solely in competitive advantage or mission outcomes. It will be measured by progress made together: toward a more trustworthy, equitable, and effective use of data and AI.

VOICES IN THE FIELD
Tim Berners-Lee on Openness

"The Web was not designed to be owned by any one person. It was designed to be a collaborative space where everyone could contribute."[11]–Tim Berners-Lee

COMMENTARY: The father of the web reminds us that openness and collaboration are not idealistic add-ons; they are the very foundations of transformative ecosystems. The same must hold true for data and AI.

11. Berners-Lee, Tim. The World Wide Web: A Very Short Personal History. World Wide Web Consortium (W3C), 1998, www.w3.org/People/Berners-Lee/ShortHistory.html.

Chapter 2

WHY DATA & AI LEADERSHIP MATTERS NOW

Organizations have reached a real inflection point: data and AI now shape outcomes at the front line. Crossing that point takes leadership. The work is stewardship, accountability, and vision that turn fragmented assets into a coherent, trusted system of advantage, embedded in everyday decisions at the highest levels.

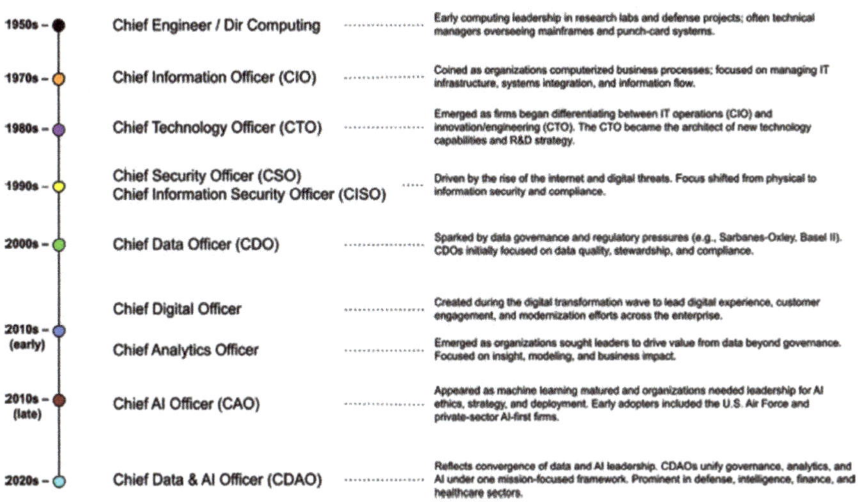

Figure 1 - The Evolution of Technical Leadership

The Conceptual Evolution of Data Leadership

To understand why data and AI leadership matters now, it is helpful to examine how far the role has evolved. The journey of data leadership is not a straight line but a series of transformations, each shaped by the technology, priorities, and cultural context of its time. What began as a custodial function buried in IT has matured into a strategic seat at the executive table.

From Custodial IT to Strategic Asset

In the 1990s and early 2000s, data primarily resided within information technology departments. Database administrators and reporting specialists focused on keeping systems available and compliant. The work was essential but rarely strategic; data was treated as a cost to manage rather than a source of insight. Early business intelligence tools offered a glimpse of possibility, allowing organizations to analyze what had happened, but the focus remained retrospective and fragmented.

The rise of the internet economy changed that equation. Companies like Google and Amazon showed that data was not exhaust from operations but fuel for innovation. Every transaction, search, and interaction became an input that refined algorithms, optimized logistics, and shaped products in real time. The arrival of big data technologies in the 2010s expanded what was possible, making it feasible to process vast, unstructured information, from social media to sensors, and revealing that the advantage would belong to those who could turn scale into usable knowledge.

As this mindset took hold, leadership followed. Forward-looking organizations began to formalize responsibility for data, recognizing that stewardship and value creation required executive accountability. The emergence of the Chief Data Officer signaled that data had finally moved from the basement of IT to the boardroom; no longer a byproduct of operations, but a strategic asset at the core of modern enterprise.

The Fusion of Disciplines

As data evolved from a back-office function to a strategic asset, the leadership required to guide it also changed. Traditional executives could not simply inherit this responsibility; the role itself became inherently hybrid.

Effective data and AI leadership now demands fluency across once-separate disciplines, blending analytical rigor, technological understanding, organizational acumen, and ethical judgment.

The rise of AI has deepened this fusion. Machine learning systems generate insights that must be explained, audited, and governed, requiring leaders who can communicate complexity clearly and earn trust across technical, regulatory, and public domains. The modern Chief Data and AI Officer is therefore as much a strategist and diplomat as a technologist.

Today, that versatility is essential. Data now defines economies, while AI shapes daily decisions, products, and policies. Societies are debating their ethics, and organizations are navigating global volatility. In this environment, strong data and AI leadership is not a luxury but a prerequisite for resilience and advantage. This turns information into responsible action at the pace modern life demands.

Recognizable Leaders Who Shaped the Field

Every field has its pioneers, and data and AI leadership is no exception. What makes this field unique is that its pioneers often had to build their own platforms, carving out new roles and reshaping executive thinking in the process. The leaders who stepped forward did more than manage datasets or champion algorithms. They shifted cultural norms, reframed what was possible, and brought urgency to issues of accountability, accessibility, and impact. Their stories illustrate not only the diversity of leadership in this space but also the outsized influence individuals can have on institutions, markets, and societies.

- **DJ Patil**: First US Chief Data Scientist who proved data leadership can move national policy, with programs like Precision Medicine and the Police Data Initiative. He set the precedent for placing data leaders at the center of public decision-making.
- **Cathy O'Neil**: Gave leaders a vocabulary and urgency for algorithmic accountability, shifting ethics from niche concern to boardroom mandate.

- **Thomas Davenport (and DJ Patil)**: Named and legitimized the data scientist role for executives, catalyzing hiring, programs, and investment.
- **Andrew Ng**: Scaled practical AI education and adoption, arguing for accessible, sector-wide use.
- **Clare Martorana**: Showed how data leadership rebuilds trust in government by modernizing services, securing data-sharing, and user-centered delivery.
- **Others in brief**: Jeff Dean advanced deep learning at scale and open ecosystems; Fei-Fei Li modeled human-centered AI and catalyzed vision breakthroughs; Jer Thorp broadened data storytelling to civic life; Anthony Goldbloom opened global collaboration through competitions.

Together, these leaders demonstrate the breadth of what data and AI leadership can mean; this list is by no means exhaustive, and continues to grow every day. Some shaped the profession, others elevated ethics, still others opened access or drove technical breakthroughs. Each, in their way, changed the face of data for the better.

The Broader Lesson

What unites all of these figures is not just their technical insight but their vision. They saw futures that others did not and were willing to push boundaries to bring them into reality. Just as importantly, they reminded the world that leadership in data and AI is not about technology alone. It is about shaping institutions, influencing culture, and earning trust. Their stories serve as both inspiration and blueprint. They show what is possible when leadership elevates data from siloed experiments to engines of transformation.

The Outsized Impact of Good Data Leadership

The transformative power of data and AI does not unfold automatically. It depends on leadership that decides how data is collected, shared, governed, and applied. Across sectors, strong leadership consistently determines whether data becomes an engine of progress or a source of friction.

In healthcare, decisive leadership around shared data platforms accelerated vaccine development during the COVID-19 pandemic, enabling global collaboration that saved lives and reshaped how science responds to crisis.

In defense and intelligence, programs such as the Department of Defense's Joint All-Domain Command and Control (JADC2) demonstrate how integrating data across domains increases decision speed and operational advantage. The result is not just efficiency but superiority in the information domain, where leadership determines who acts faster and with greater confidence.

In the private sector, Amazon's leaders treated every process as data to be modeled and optimized, turning a retailer into a predictive enterprise. Netflix used viewing data to personalize recommendations and guide content investment, redefining what creative industries could achieve through analytics.

Finance offers its own lesson. After the 2008 financial crisis, forward-looking data leaders turned regulatory compliance into a modernization effort, embedding governance and AI-driven risk management across operations. In recent market shocks, those same practices enabled near-real-time risk modeling and portfolio adjustments—proof that disciplined data leadership builds both trust and resilience.

Across all these domains, the pattern is clear: when leadership elevates data and AI from technical tools to strategic imperatives, the results multiply; lives are saved, missions secured, markets transformed, and institutions strengthened. Where leadership is absent, fragmentation, mistrust, and stagnation take hold.

What Data and AI Leadership Brings to the C-Suite

To thrive today, data and AI leadership must sit at the executive table, not buried in technical silos. When data leaders are empowered at the C-suite level, they bring a capability that reshapes how organizations think and act across three dimensions: strategic clarity, operational discipline, and innovative leverage.

Strategic Clarity

Data leaders turn information into insight that decision-makers can trust. By designing systems that emphasize accuracy, timeliness, and context, they ensure decisions are grounded in evidence rather than intuition. During the pandemic, cities with chief data officers could rely on near-real-time information on hospital capacity and infection rates, enabling faster and more targeted public health responses.

Operational Discipline

Leadership also brings structure to how data is governed. Strong data executives enforce standards, maintain lineage transparency, and balance accessibility with accountability. In financial services, this discipline enabled institutions to meet stringent regulatory requirements and transform compliance into a competitive advantage, demonstrating that stewardship and agility can coexist.

Innovative Leverage

Finally, data and AI leaders channel innovation toward measurable value. They separate signal from hype, align experiments with strategy, and scale what works. In healthcare, for instance, empowered analytics leaders moved beyond pilots to deploy AI triage tools that meaningfully reduced patient wait times, demonstrating how responsible innovation can deliver tangible outcomes.

Taken together, these capabilities make data and AI leadership indispensable. Without them, organizations drift into fragmented efforts, duplicated systems, and eroding trust. With them, they gain coherence, confidence, and a durable advantage in how they learn, decide, and evolve.

Trust as the Invisible Currency

Trust is the invisible currency of the data and AI era. It rarely appears on a balance sheet, yet it underpins every meaningful outcome. Without trust, even the most advanced systems fail to gain adoption; with it, modest capabilities can achieve extraordinary results. Building and sustaining that trust is one of the central responsibilities of data and AI leadership.

Trust Within Teams

Trust begins close to home, among the teams that collect, curate, and apply data. When people doubt the quality of the information they work with or suspect that leaders are shaping outcomes to fit a narrative, collaboration quickly unravels. Strong leaders prevent this by investing in data quality at the source, establishing clear lineage so everyone knows where data comes from, and ensuring analyses are reproducible. When those foundations are in place, confidence grows. When they are not, teams hedge their conclusions, decisions slow, and innovation stalls. The absence of internal trust becomes a silent tax on progress.

Trust Across the Organization

Beyond the team, trust depends on how data leaders bridge the gap between technical expertise and executive decision-making. Transparency and auditability are the twin anchors of credibility. When executives understand how data was sourced, the limitations that exist, and how the results were derived, they act with confidence. When a model denies a loan or flags a transaction, leaders should be able to trace that decision through its inputs and logic. Such traceability turns data-driven decisions from mysteries into accountable processes. The opposite is equally true: a single faulty metric or unexplained algorithmic outcome can erode confidence for years, forcing leaders to fight uphill battles to restore belief.

Trust Beyond the Organization

The outer circle of trust encompasses the community, comprising customers, citizens, partners, and regulators. Here, the stakes are highest because once broken, trust is slow to return. Businesses earn it through openness: explaining how data is used, offering real consent, and communicating the value exchanged. Companies that obscure their practices may achieve short-term gains, but they inevitably face a reputational collapse. In the public sector, leaders who champion open data programs demonstrate how visibility can enhance trust in institutions, enabling citizens to see and verify how decisions are made. Regulators, too, reward openness; organizations that demonstrate traceability often gain greater freedom to innovate, while those that resist scrutiny invite constraint.

When Trust Breaks

Trust can take years to build but only moments to lose. A flawed credit model that denies legitimate applicants or a recruiting algorithm that encodes bias can spark lasting damage. Customers question fairness, regulators intervene, employees disengage, and the organization's momentum grinds to a halt. In effect, it loses its license to operate in the data and AI space.

The Leadership Imperative

For data and AI leaders, trust must be designed, not assumed. It is built through quality, sustained through transparency, reinforced through auditability, and expanded through accountability. Leaders who embody these principles move faster, collaborate more openly, and earn the confidence to innovate; those who do not soon discover that no amount of technology can substitute for belief. In the era of data and AI, trust remains the currency that enables every other transaction.

Sectoral Contrasts: Leadership in Different Arenas

While the responsibilities of data and AI leaders (building trust, ensuring quality, and guiding innovation) are universal, their expression varies across domains. Each sector faces unique pressures shaped by its mission, regulations, and public expectations.

In **defense and intelligence**, leadership is measured in speed, integration, and security. The challenge is not efficiency but mission success: connecting data across land, air, sea, space, and cyber domains to enable real-time decision-making. A single failure of coordination can turn information advantage into vulnerability.

In **healthcare**, leadership strikes a balance between innovation and ethics. Data and AI promise breakthroughs in treatment and prediction, but lives are at stake. Leaders must safeguard privacy, validate results, and earn the trust of clinicians and patients alike, proving that technology can advance care without sacrificing dignity or safety.

In **the consumer and retail sectors**, leadership defines the line between personalization and intrusion. Companies that use data to anticipate needs and respect privacy tend to thrive; those that cross ethical boundaries often face backlash. Success here depends on reading cultural shifts as closely as market trends.

In **finance**, leadership is about compliance and resilience. Institutions must govern data transparently, model risk in real time, and maintain public and regulatory trust even amid shocks. A failure of data stewardship can have a ripple effect across entire economies.

Across all sectors, the pattern is clear: strong data and AI leadership transforms data into mission advantage, trust, and growth; weak leadership turns it into exposure and risk.

Chapter 3

THE CATALOG FRAMEWORK

The work of a Chief Data and AI Officer, or any leader charged with integrating data and AI into mission and enterprise, rarely begins in a pristine environment. You inherit a living system shaped by legacy decisions, embedded cultures, urgent pressures, and uneven capabilities. Progress seldom comes from a single breakthrough or a silver-bullet tool. It comes from orchestrating a set of disciplines, each essential and interdependent.

The CATALOG Framework offers that orchestration. It is not a checklist or rigid doctrine but a map that ensures no vital aspect of transformation is neglected in the rush to deliver. Using the mnemonic "CATALOG", it names seven domains that, cultivated together, form the backbone of effective data and AI leadership:

1. **C**ulture & Talent
2. **A**nalytics & AI
3. **T**echnology & Architecture
4. **A**lignment (Strategy & Policy)
5. **L**eadership & Governance
6. **O**perations & Delivery
7. **G**rowth & Measurement

Each of these domains represents a discipline in its own right. Taken together, they form the comprehensive field of action for leaders charged with unlocking the promise of data and AI. What follows is not a surface overview, but a closer study of each; why they matter, how they can be activated, and what enduring success looks like in practice.

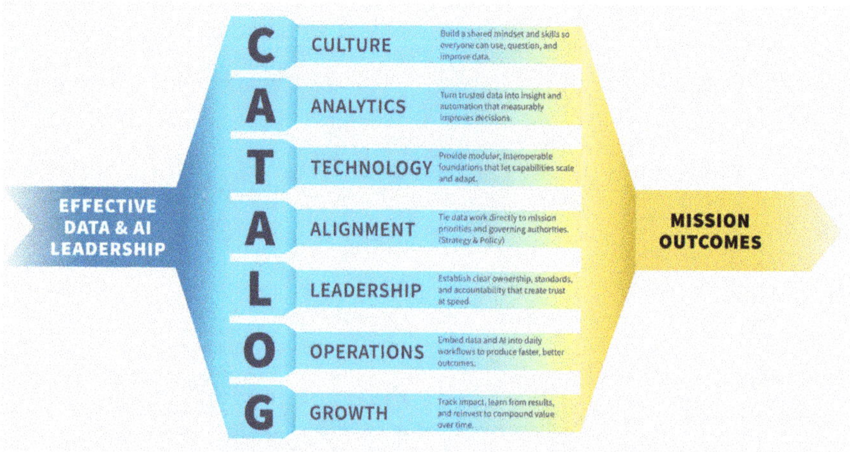

Figure 2 - 7 Domains of the CATALOG Framework

1. Culture & Talent

Systems enable. People sustain. A culture that treats data as someone else's problem cannot metabolize even the best architecture. A workforce that prizes stewardship, curiosity, and evidence wrings value from imperfect tools and makes better ones inevitable.

Culture reveals itself in small moments. A senior leader asks for the data behind a claim and listens when it challenges a favored intuition. A frontline team experiments with a dataset and is celebrated for the attempt rather than punished for imperfection. A steward declines an access request that violates policy and is backed by leadership. Each of these teaches the organization what is serious.

Edgar Schein, whose work on organizational culture has guided leaders for decades,[12] taught that culture operates at three distinct levels. At

12. Schein, Edgar H. Organizational Culture and Leadership. 5th ed., Jossey-Bass, 2016.

the surface are artifacts: the visible behaviors, dress codes, and physical arrangements that anyone can observe. Beneath that are espoused values: the stated principles and strategies that organizations claim to follow. But at the deepest level are basic underlying assumptions: the unconscious, taken-for-granted beliefs that actually determine behavior.

This framework explains why so many data transformation efforts fail. Organizations invest heavily at the artifacts level, buying dashboards and creating data science teams, while leaving the deeper assumptions untouched. They articulate espoused values, declaring themselves "data-driven" or "analytically rigorous," but the underlying assumptions remain unchanged: that decisions are really made by the most senior person in the room, that intuition trumps evidence, that raising uncomfortable data is career-limiting.

Real culture change requires working at all three levels simultaneously. The artifacts matter: visible data leaders, accessible tools, and physical spaces where analysts and operators collaborate signal what the organization values. Espoused values matter: executives must articulate why data and evidence are central to the mission, not just nice-to-haves. But unless the basic assumptions shift, the first two levels remain theater. The real work happens when leaders consistently act in ways that contradict old assumptions and reinforce new ones: promoting people who challenged intuition with data, rewarding teams who surfaced inconvenient truths, backing stewards who said no to powerful requesters.

Literacy matters, but training alone does not shift culture. People must see a future for themselves in the data-driven world. Visible career paths for data talent, rotations between operations and analysis, and communities of practice that welcome naïve questions turn aspiration into habit. Incentives do quiet work. When promotions and praise accrue to outcomes demonstrably improved by data, norms begin to shift.

Beware slogans without practice. "Data-driven" means little if the first time data contradicts a hunch, the numbers are blamed. "Fail fast" is hollow when a small failure becomes a career setback. Trust grows when words, rewards, and consequences align.

When culture turns, you can feel it. Meetings begin with what we know, then move to what we suspect. Teams ask each other for data because it is polite. Pride in stewardship rises, and impatience with untestable claims becomes healthy pressure. Governance becomes easier, architecture improves, and AI finds its place.

VOICES IN THE FIELD
Federal CDO Expectations

"I thought I'd spend my first year buying tools. I spent it learning to speak strategy and policy with fluency. Once leadership heard their priorities reflected in our work, everything else got easier."
–Senior CDO, U.S. federal agency

2. Analytics & AI

This domain changes the tempo of work. Properly applied, analytics and AI enable organizations to move from hindsight to foresight, from manual toil to automated reliability, and from reactive cycles to a proactive posture. The danger lies in spectacle, in demos that impress but never integrate.

Begin where pain and measurability intersect. Which workflows drown teams in repetitive tasks? Which decisions suffer from latency or blind spots? Which outcomes are already tracked so that impact can be proven? A modest automation that returns thousands of hours is preferable to a brilliant prototype that requires caveats and a bespoke environment. Fit to context is the difference between adoption and shelfware.

Treat explainability as a feature, not an afterthought. A model that cannot be interrogated will not survive oversight or a bad outcome. Build monitoring and feedback loops from the start. The first version is the worst it will ever be, and improvement must be a daily habit.

Scaling requires choreography. A pilot should launch with a clear path to production that defines where it will run, how it will be secured, and who will own and support it. Funding should anticipate that success creates demand. Governance must mature alongside capability, with registries, documented assumptions, bias testing, and performance SLAs. These are how you earn the right to place machines near consequential decisions.

When AI lands, it becomes ordinary in the best sense. Leaders stop asking "Where is the AI?" and start asking "What did we learn?" Analysts reconcile less and reason more. Operators say a tool helped them do their job. That is success.

QUICK NUMBERS
The Cost of Poor Data

Gartner has estimated that poor data quality costs organizations an average of $12.9M annually[13] through rework, missed opportunities, and compliance risk. Governance is not overhead; it's a margin and mission defense.

3. Technology & Architecture Modernization

Modernization is not a shopping trip. It is a commitment to adaptability. Brilliant analytical ideas are stifled in brittle, proprietary mazes. Thoughtful architecture, designed to be modular, standards-forward, and transparent, turns modest tools into a platform for compounding advantage.

Trace how work actually gets done. Which decisions need which systems, and how does data arrive? Watch the seams: CSVs by email, scripts on contractor laptops, duplicate warehouses because no one trusts the enterprise one. Those seams define what your architecture must solve.

Early wins are connective. Expose a high-value dataset through a secure API so teams stop scraping. Consolidate duplicative reporting into a single curated feed and retire the rest. Pilot a lightweight data fabric that stitches sources together for one operational use case and prove that federation can deliver coherence without forced centralization. Measure the time saved and the decisions accelerated. Architecture earns legitimacy when it reduces toil.

At scale, insist on disciplines such as versioned data contracts to prevent surprise, guardrails that make the paved road the easy one, reference patterns for self-service, and financial transparency so leaders understand cost. Modernization is less a moonshot than a rhythm.

13. Gartner. Gartner Research Identifies Data Quality as an Enterprise-Wide Issue Costing Organizations an Average of $12.9 Million Annually. Gartner, 6 May 2021, https://www.gartner.com/en/newsroom/press-releases/2021-05-06-gartner-research-identifies-data-quality-as-an-enterprise-wide-issue-costing-organizations-an-average-of--12-9-million-annually

When it takes root, experimentation no longer threatens stability. New partners plug in without heroics. A model proven in one unit is adopted by another with minimal ceremony. The system feels less like a museum of past decisions and more like a workshop ready for the next one.

VOICES IN THE FIELD
Modular by Design

"We stopped hunting for the mythical 'single system.' We invested in the connective tissue: identity, APIs, catalog, observability. Suddenly our product and mission teams could ship. We didn't predict the future; we designed for it."
–CTO, Fortune 500 defense contractor

4. Alignment (Strategy & Policy)

Enduring transformation begins with a theory of success, and in public institutions and regulated industries, that theory is inseparable from policy. Strategy defines purpose and desired outcomes. Policy confers authority, sets priorities, and allocates resources. Alignment translates those imperatives into practical work, making data and AI the most direct path to declared goals.

Start by learning the organization's language. Read strategies, budget justifications, commander's guidance, and executive memos. Listen for recurring outcomes such as readiness, trust, efficiency, and resilience. Map them against governing rules like executive orders, audits, and internal directives. You are not cataloging paperwork; you are defining the boundaries of legitimate action.

From there, alignment becomes narrative craft. A data-sharing initiative is not "an API program." It is how decision cycles compress from weeks to days. A catalog is not "metadata hygiene." It is how audit burden falls, and trust rises. Speak this way and your portfolio moves from optional to inevitable.

Create a compact alignment map early. On one side, list the outcomes promised. On the other, show the data and AI capabilities that make them possible. The power of this artifact lies in its clarity: everyone can see how initiatives trace back to strategy and rest on policy.

Alignment has a rhythm. Strategies evolve, leaders rotate, and crises shift priorities. Treat the map as living. The mature signal is not applause but access: being invited into strategy discussions before decisions are finalized, because colleagues recognize that data and AI determine what is possible.

When alignment works, it feels like gravity. Funding paths open, oversight questions ease, and leaders cite your programs in their own briefings. You are no longer a petitioning technologist. You are an author of the operating theory.

TREND SNAPSHOT
Scaling Beyond Pilots

McKinsey's *State of AI* (McKinsey 2023[14]) found **only ~23% of organizations have moved beyond pilots to widespread adoption**. The throughline among those that scale: explicit ties to strategy, strong data governance, and architectures that make productionization routine rather than heroic.

5. Leadership & Governance

If alignment is the compass, governance is the terrain, and the terrain determines how far the journey can go. Governance is the architecture of trust. It is the disciplined work that makes data findable, comprehensible, protected, and fit for purpose.

You can often learn more about governance by tracing a single metric than by reading a thick policy. Where is the authoritative source, and who owns it? If the answer varies, that reveals trust gaps, not just technical ones. Who holds stewardship and with what authority? Titles without power cannot enforce definitions or access.

Prove that governance accelerates by repairing one visible, high-value asset end-to-end. Designate a steward empowered to say yes and no. Make lineage audible. Capture meaning in the language people use. Enforce

14. McKinsey & Company. The State of AI in 2023: Generative AI's Breakout Year. McKinsey & Company, 6 Dec. 2023, https://www.mckinsey.com/capabilities/quantumblack/our-insights/the-state-of-ai-in-2023-generative-ais-breakout-year

access through identity, not spreadsheets. Show that the reconciliation time decreases and the escalation volume declines.

At scale, governance must evolve from procedure to reflex. Definitions live in the catalog, not in PDFs. Quality checks run in pipelines. Access controls exist as code. Stewardship becomes distributed, with credibility strengthened when every unit has a named steward accountable to peers.

Governance fails when it becomes ornamental or aspirational. Anchor it to pain such as audits, public releases, or cross-unit reporting, and let relief prove its worth. When governance works, trust becomes ambient. Systems meet obligations through design rather than heroics. Governance is not the opposite of agility. It is what allows agility to scale.

6. Operations & Delivery

Everything culminates in the doing. A capability matters when it changes how the mission progresses: faster cycles, better allocation, fewer errors, and greater resilience. Integration embeds data and AI so deeply into daily work that people stop noticing the technology and start seeing the results.

Proximity is key. Embed data talent where decisions are made. Sit with operators. Many "requirements" written at a distance dissolve when you see the actual tempo and constraints. Solutions that fit come from empathy: a display that shows what matters now, a model that favors explainability when a human must justify a choice, and an alert that is rare enough to be trusted.

Integrate with existing rhythms. If there is a daily operational brief, produce artifacts for it. Respect existing chains of authority. Measure value in the same units leaders already use, such as days saved, claims resolved, patients seen, or sorties generated.

Sustainment is part of success. Budget for maintenance. Train backups. Rotate stewards. When a tool becomes indispensable, reliability becomes mission-critical.

Integration falters when technology is pushed like a foreign object, or when success is measured by models built rather than outcomes changed. The countermeasure is co-ownership. Users feel agency over evolution. Builders remain reachable. The capability becomes how we work now.

When integration sticks, decision cycles compress, and resilience rises.

"Next week" becomes "this afternoon." In a crisis, data and tools are where they need to be, when they need to be there.

7. Growth & Measurement

Transformation endures when it learns. Growth and measurement turn progress into momentum. Too often, teams perform mental gymnastics to either (a) get the measurements to fit the narrative or (b) only measure things that will show positive growth and cast the team in the best light.

Define metrics that matter, such as time to insight, reuse of governed assets, cycle-time reductions, improvements in readiness, and resolved audit findings. Track them openly. Celebrate measurable gains, but also study what lagged. Use retrospectives to adjust governance, architecture, and incentives.

Growth fails when it becomes performative (e.g., counting dashboards rather than measuring impact), and succeeds when measurement fuels adaptation.

When growth becomes habit, innovation compounds. Teams anticipate what leaders will need, and learning becomes part of the operating rhythm. Progress is no longer episodic; it is continuous.

QUICK NUMBERS
Decision Tempo

A 2022 GAO review reported that agencies embedding analytics into daily workflows saw **30–50% reductions in decision cycle times**. Value compounds when the insight is present at the moment of choice, not days later in a slide deck (GAO, 2022[15]).

15. United States Government Accountability Office. Data and Analytics Innovation: Agencies Increased Use of Data and Analytics to Improve Decision-Making. GAO-22-104666, U.S. Government Accountability Office, Mar. 2022.

The Fabric of the Framework

None of these domains is sufficient on its own, and none remains stable if the others falter. Alignment earns the right to build. Governance makes the substrate trustworthy. Architecture keeps you adaptable. Analytics and AI move the needle. Culture sustains the behavior. Operations turn potential into value. Growth ensures learning and renewal.

Work any one in isolation and you will hit limits quickly. Work them together, deliberately and iteratively, and you create a system that learns.

Chapter 4

ORGANIZATIONAL DESIGN & CHANGE LEADERSHIP

The first several chapters of this book lay the essential foundations for data and AI leadership, including the strategic alignment of missions and policies, the cultural and technical challenges that shape transformation, and the frameworks that help leaders navigate this evolving field. With those building blocks in place, this chapter marks a turning point. It moves from ideas to application, from concepts to operations.

Organizational design and change leadership represent the first layer of practice where theory meets reality. No matter how sound a strategy may be, it will falter without structures that support it and leaders who can guide people through transition. This is where data leaders like the CDAO begin to translate vision into action. The concepts described earlier: policy alignment, decision tempo, and socio-technical framing, come to life here as living systems within teams, portfolios, and enterprises.

In this chapter, we explore how to design organizations around the decisions that matter most, how to select and adapt archetypes of organizational form, how to clarify the roles and responsibilities that so often create confusion, and how to lead change in a way that makes transformation durable. We also address the discipline of measuring what matters and the challenge of moving from short-lived programs to enduring institutions. Each of these topics builds on the foundation laid in earlier chapters, but they also set the stage for what comes next. Later sections of the book will extend beyond internal structures to consider external ecosystems, ethical guardrails, and the cultivation of talent.

Readers should approach this chapter as both a guide and a bridge. It offers operational techniques for immediate use while also laying the groundwork for subsequent chapters that expand into ecosystems, risk, trust, and long-term culture. By the end of this chapter, the reader should see not only how to organize for near-term decision advantage, but also how these organizational patterns can grow into the lasting resilience that the next era of data and AI leadership demands.

Designing Around Decisions

Organizational design anchored in data and AI transformation cannot be reduced to charts, reporting lines, or platform diagrams. At its core, the principle of designing around decisions means that every structural choice should be made with a clear understanding of the mission outcomes it is intended to serve. The mission defines what decisions must be made, at what speed, with what information, and under what conditions of uncertainty or risk. The role of organizational design is to ensure that people, teams, processes, and technologies are configured to enable more effective and confident decision-making.

In practice, this requires starting not with the data or the systems, but with the questions operators and leaders need answered. For a defense mission, this may be how to prioritize limited ISR (intelligence, surveillance, and reconnaissance) assets across multiple theaters in real-time. For a healthcare mission, it could involve identifying emerging patterns in patient data that signal the need for rapid intervention. In the private sector, the same logic applies: logistics companies have restructured supply chain teams around critical decisions about routing and inventory allocation. In each case, the organization's structure reflects the pressure points where better decisions yield disproportionate value.

The experience of successful organizations reinforces this lesson. The United States Special Operations Command built joint task forces that integrate operators, intelligence analysts, technologists, and logisticians into a single decision cell. The structure was deliberately designed to collapse layers of bureaucracy, allowing time-sensitive targeting decisions to be made within minutes rather than hours. In the private sector, Amazon's famed "two pizza teams" follow the same pattern. Small, empowered, cross-functional groups own end-to-end decisions about products or services, supported by common data and infrastructure platforms. Their autonomy is not a matter of culture alone but is structurally embedded to ensure decision speed and accountability.

Designing around decisions also means clarifying what level of decision-making belongs at each tier of the organization. Strategic choices about mission priorities may rest with senior leadership, while operational allocations and tactical adjustments are delegated to product teams with direct access to data and operators. The structure should be flexible enough to allow information to flow upward when emerging risks or opportunities demand visibility, yet also robust enough to enable downward delegation without compromising control. Striking this balance ensures that decision-making occurs at the lowest level capable of acting, thereby improving speed and enhancing resilience.

An essential feature of this approach is that platforms and governance functions exist to serve decisions rather than to dictate them. The data catalog, the semantic layer, or the access control system are valuable only insofar as they make it easier and safer for teams to deliver decision-support products. This echoes practices in high-performing firms such as Capital One, which restructured its technology and risk functions so that compliance and control are automated within platforms rather than imposed through manual reviews. The effect is to shift governance from a bottleneck to an enabler, one that accelerates decision cycles rather than slowing them down.

Ultimately, designing around decisions requires leaders to visualize the decision chain as the backbone of the organization. From sensing to analysis to choice to action, the structure should mirror that flow. Teams should be situated where they can shorten the chain, platforms should be built to support each link, and governance should provide assurance without delay. When structure and decision outcomes align in this way, organizations consistently demonstrate not only improved operational performance but also greater adaptability to change.

Archetypes of Organizational Form

Organizations rarely begin with a blank slate. They inherit structures, authorities, and cultures that have often evolved over decades. For a CDAO, the challenge is not to erase those realities but to shape them into forms that enable data-driven decisions. Four archetypes provide different pathways for doing so. Each carries strengths and tradeoffs, and each can be operationalized in concrete ways.

The need for multiple organizational forms reflects a fundamental tension that Clayton Christensen identified in "The Innovator's Dilemma." Christensen demonstrated why successful, well-managed companies often fail when confronted with disruptive innovation.[16] The very practices that make organizations efficient at executing their current business model (rigorous planning, stage-gate approvals, resource allocation tied to proven markets) become obstacles when pursuing innovations that don't fit existing frameworks. Established processes optimize for sustaining innovation that serves existing customers with proven business models, but they systematically reject disruptive innovations that target new markets or create new value propositions.

For data and AI leaders, Christensen's insight is critical: you cannot transform an organization using only the structures and processes that made it successful in the pre-AI era. The central IT organization that excels at maintaining enterprise systems and ensuring compliance will struggle to keep pace with the speed required for AI experimentation. The budget process designed to fund multi-year infrastructure projects will starve initiatives that need to iterate weekly. The governance frameworks built for predictable, well-understood technologies will delay deployments of capabilities that are still being defined.

Christensen argued that organizations facing disruptive innovation need to create separate organizational structures with different processes, cost structures, and metrics. This is precisely what the skunkworks archetype provides: a protected space where teams can operate outside the constraints of the mainstream organization, move quickly, take risks, and prove concepts before they must face the full weight of enterprise processes.

16. Christensen, Clayton M. The Innovator's Dilemma: When New Technologies Cause Great Firms to Fail. Harvard Business School Press, 1997.

But Christensen also warned that if successful innovations remain permanently separated, they never transform the core business. The challenge is not just to create skunkworks but to build pathways for reintegrating successful innovations back into the organization.

This organizational challenge was also recognized by Tom Peters and Robert Waterman in "In Search of Excellence," where they documented Lockheed's famous Skunk Works as a model for innovation within large organizations.[17] Led by Kelly Johnson, Lockheed's Skunk Works operated with remarkable autonomy, minimal bureaucracy, and direct access to senior leadership. The team delivered breakthrough aircraft like the U-2 and SR-71 by insulating engineers from corporate processes that would have slowed them down. Peters and Waterman identified this pattern across excellent companies: they maintained "skunk works" or similar structures to pursue innovation while their mainstream operations focused on execution.

The four archetypes that follow represent different strategies for managing this tension. The central platform with embedded pods attempts to build innovation capacity within existing structures. The hub-and-spoke model creates flexibility while maintaining coherence. The federated product lines acknowledge that some missions require full autonomy. And the skunkworks-to-scale approach explicitly creates a separate space for innovation with a planned pathway back to the mainstream. Each reflects a different answer to Christensen's dilemma: how to pursue disruptive innovation without destroying the capabilities that make the core organization effective.

The first archetype is the **central platform with embedded mission pods**. Here, a relatively small but highly skilled central team maintains the common platform, curates governance standards, and ensures compliance with policies and regulations. Around that platform, mission pods are embedded with operators to deliver products tied directly to mission outcomes. This approach is especially valuable in the early stages of transformation when skills are scarce, and consistency is vital. Operationalization involves establishing a thin but resilient platform layer that provides cataloging, lineage, and secure workspaces, while rotating small

17. Peters, Thomas J., and Robert H. Waterman Jr. In Search of Excellence: Lessons from America's Best-Run Companies. Harper & Row, 1982.

cross-functional pods into mission areas for short but impactful engagements. Indicators of success include a rapid reduction in time to deploy first products, a clear sense among operators that they can trust the guardrails, and a platform backlog that reflects real mission demand rather than abstract technical priorities.

The second archetype is the **hub-and-spoke model**, often described as a community of practice. In this structure, the central hub defines golden paths, standards, and reusable assets, while the spokes comprise mission teams that adapt and extend those assets to meet their local needs. This model is well-suited to environments where missions have some existing capacity in data and AI but benefit from alignment around shared practices. To operationalize it, leaders establish a central certification program, maintain repositories of pre-approved modules, and run regular cross-team architecture reviews. Success is evident when mission teams independently deliver products without waiting for central approval, when deviations from standards are intentional and well-documented rather than accidental, and when the hub's role shifts from control to enablement. During implementation, evidence of healthy adoption is evident in shared code contributions, common vocabulary across different units, and a measurable decline in duplicated efforts.

The third archetype is **mission federated product lines**. This structure acknowledges that some missions are so diverse in their requirements and environments that centralization cannot deliver sufficient agility. For example, intelligence missions operating at different classification levels may require unique enclaves that cannot be easily standardized. In such cases, federated teams own full product lines from data ingestion through to delivery, but they remain linked through contractual interfaces such as APIs and ontologies. Operationalizing this model requires establishing clear interoperability rules, shared semantic standards, and internal source modules that can be reused across enclaves. Indicators of success include the ability to deploy mission-specific products quickly without lengthy exception processes, a gradual convergence of semantic standards across the enterprise, and evidence that federated teams are contributing improvements back into the common ecosystem rather than reinventing in isolation.

The final archetype is the **skunkworks-to-scale approach**. This begins with small, empowered teams that operate with significant autonomy to deliver highly visible, quick wins. The purpose is to generate momentum and proof of value in contexts where skepticism or inertia may otherwise block progress. Operationalizing this model requires clarity about exit criteria. Skunkworks projects must be designed with a pathway to production from the start, including early involvement of compliance and security authorities. Success is indicated first by the ability to demonstrate a compelling prototype that captures leadership attention, and later by the smooth transition of that prototype into operational environments without accruing crippling technical or policy debt. Over time, the skunkworks mentality should give way to institutionalization, with practices and lessons absorbed into the mainstream portfolio.

These archetypes are not mutually exclusive. An organization may begin with a central platform and embedded pods, mature into a hub-and-spoke model, and later allow specific enclaves to operate as federated product lines. At different points, skunkworks projects may inject energy and innovation into the cycle. What matters is that leaders consciously select and adapt these forms to match their mission context. Before implementation, indicators such as the availability of skilled personnel, the diversity of mission needs, and the maturity of existing governance should guide the choice. During implementation, measures of adoption, trust, and speed provide feedback. Afterward, sustainability is the key indicator, seen in whether decision-making has genuinely accelerated, whether controls are embedded seamlessly, and whether the structure continues to evolve as missions and technologies change.

TREND SNAPSHOT
Shifts in Data Organization Models

Research shows evolving preferences in how organizations structure their data and analytics functions:

- A study of 100 data and analytics leaders found **42%** said the **hub-and-spoke model** was ideal for their business, compared to **38%** favoring decentralized models and just **20%** preferring fully centralized operations. This indicates a strong trend toward hybrid structures that balance control with agility.[18]

- Additional data from a survey of 300 executives revealed that **21%** operate with a formal hub-and-spoke setup, while **52%** have a coordinated or federated structure. Only around a quarter remains fully centralized.[19]

These numbers reflect a growing recognition that neither extremes of full centralization nor full decentralization consistently deliver on both alignment and agility. Hybrid models like hub-and-spoke are gaining favor as they allow organizations to evolve structures in step with capacity, mission diversity, and technological maturity.

Roles, Responsibilities, and Decision Rights

When organizations embark on data and AI transformation, one of the most persistent challenges is clarifying who is responsible for what. Titles have proliferated in recent years. Chief Data Officers, Chief Analytics Officers, Chief Artificial Intelligence Officers, and the emerging Chief Data and AI Officer role now exist alongside traditional technology leaders such as the Chief Information Officer and Chief Technology Officer. Below them, Vice Presidents of Data Engineering, Directors of Analytics, Product Managers, and Stewards all hold essential responsibilities. Without deliberate design, these roles can overlap or conflict, leading to confusion and diluted accountability. A mature data and AI function resolves this by clearly mapping responsibilities and decision rights.

The Chief Information Officer has historically held responsibility for core IT infrastructure, enterprise systems, and overall technology spending. In many organizations, the CIO remains a central figure; however, the rise of data-intensive operations has necessitated a redistribution of authority. The Chief Technology Officer often provides a forward-looking technical

18. Ironside Group. "Learn Why So Many Companies Are Turning to a Hub and Spoke Data Model." Ironside Group Blog, https://www.ironsidegroup.com/blog/learn-why-so-many-companies-are-turning-toa-hub-and-spoke-data-model/ .
19. Gartner. One Minute Insight: Data and Analytics Operating Models. Gartner Peer Community, https://www.gartner.com/peer-community/oneminuteinsights/omi-data-analytics-operating-models-lp3 .

strategy and evaluates emerging platforms. In commercial enterprises, the CTO may lead innovation labs or advanced engineering teams that test new technologies. In government agencies, the CIO and CTO roles are often divided between operations and modernization, with the CIO ensuring compliance and availability while the CTO sponsors experimentation.

The Chief Data Officer emerged as organizations recognized that data required its own leadership, distinct from that of infrastructure. The CDO traditionally focuses on governance, data quality, cataloging, and stewardship. This role ensures that data can be trusted, discovered, and reused. The Chief Analytics Officer, by contrast, has been associated with advanced analytics, business intelligence, and, more recently, machine learning. Many organizations have found overlap between the two and have created hybrid arrangements. The latest evolution is the Chief Data and AI Officer, which unites stewardship of data assets with oversight of applied AI and advanced analytics. This unification is increasingly necessary because models and AI capabilities cannot be responsibly deployed without integrated data governance, and governance is incomplete without consideration of algorithmic risk.

Beneath these executive roles, operational leadership carries significant weight. A Vice President of Data Engineering or a Director of Data Platforms typically manages the teams that build and maintain ingestion pipelines, data lakes, warehouses, and catalogs. Their focus is on reliability, scalability, and performance of infrastructure. Directors of Analytics or Applied Data Science lead teams that develop models, dashboards, and decision-support products. They act as the bridge between mission needs and technical implementation. Data Stewards and Governance Leads translate policy into practical rules for classification, lineage, and access control. Product Owners and Product Managers for data and AI offerings ensure that solutions are built around user needs and mission outcomes rather than technology for its own sake.

When roles are clearly defined, organizations can disambiguate areas of focus. The CIO should retain accountability for enterprise IT operations and compliance with federal or industry regulations. The CTO should be measured on technology innovation and the ability to bring new capabilities into maturity. The CDO should be responsible for data as a trusted enterprise asset, ensuring that governance, quality, and discoverability are embedded. The CAO or analytics leader should own the development of insights and predictive capabilities. The CDAO, when established, provides the integrating force, accountable for the portfolio of data and AI products that deliver measurable decision advantage. In this model, Vice Presidents and Directors have clearer charters: data engineering leaders focus on platforms, analytics leaders on use cases, and governance leaders on translating policy to code.

Industry experience offers prescriptive lessons. At financial services firms such as JPMorgan Chase, the CDO function was initially fragmented, with separate teams for governance, reporting, and AI research. Over time, the organization aligned these under a single executive who could balance compliance with innovation. This shift enabled faster delivery of AI products because data quality and regulatory controls were no longer afterthoughts but were integrated into product design. In the public sector, agencies that have appointed CDAOs are beginning to formalize similar alignments, ensuring that model risk, data privacy, and mission analytics are addressed within a unified strategy.

The transition from today's often fragmented roles to a fully evolved model requires deliberate staging. Early steps include clarifying charters for existing executives, creating cross-functional councils to resolve overlaps, and piloting integrated teams spanning governance, engineering, and analytics. As organizations mature, roles can be consolidated, with the CDAO emerging as a natural focal point. Success indicators include reduced duplication of effort, faster product release cycles with embedded compliance, and greater clarity among staff about whom to approach for specific issues. Ultimately, the goal is to ensure that each role is empowered to focus on its unique contribution, while the collective leadership aligns around mission decisions and outcomes.

Leading Change in Practice

Designing a sound organizational structure is only part of the transformation journey. The greater challenge often lies in leading change so that structures, technologies, and policies are not only adopted but also fully integrated into the organization's culture. Change leadership is not a one-time announcement or a communications campaign. It is an ongoing process that requires clarity of purpose, deliberate action sequencing, and continuous reinforcement. For the CDAO, this responsibility sits at the heart of building a lasting decision advantage.

John Kotter's research on organizational transformation, synthesized in his influential framework for leading change, offers a disciplined approach to what otherwise feels overwhelming.[20] After studying over 100 organizations attempting major transformations, Kotter identified why most efforts fail: they skip essential steps, declare victory too early, or fail to anchor new approaches in culture. His eight-step process, while originally developed for commercial enterprises, applies directly to data and AI transformation in any context.

Kotter's framework begins with creating urgency, ensuring that stakeholders understand not just that change is coming but why delay is dangerous. For a CDAO, this means translating abstract concepts like "data-driven decision-making" into concrete stakes: the adversary who moves faster because their data integration is superior, the customer who defects because analytics are too slow, the regulator who intervenes because governance is reactive rather than embedded. Urgency cannot be manufactured through fear alone; it must be grounded in a reality that people can see and feel.

The second step is building a guiding coalition, a group with enough power and credibility to lead the effort. For data transformation, this means assembling not just technical experts but mission operators, compliance officers, and business leaders who collectively represent the organization's critical functions. Too many CDAOs attempt transformation with only their direct reports, discovering too late that real authority resided elsewhere. The coalition must be visible, aligned, and demonstrably committed.

20. Kotter, John P. Leading Change. Harvard Business School Press, 1996.

Kotter's framework then calls for developing a clear vision and strategy, communicating that vision relentlessly, empowering broad-based action by removing obstacles, generating short-term wins to maintain momentum, consolidating gains to drive deeper change, and finally anchoring new approaches in culture so they outlast individual leaders. Each step builds on the previous one. Skipping steps, or executing them superficially, is why so many transformation programs stall after initial enthusiasm fades.

What makes Kotter's framework particularly valuable for data and AI leaders is its emphasis on sequence and sustainability. It is not enough to launch initiatives; they must be sequenced to build credibility and remove blockers progressively. It is not enough to achieve early wins; those wins must be leveraged to tackle harder problems. And critically, it is not enough to change systems and processes; the underlying culture must shift, or the organization will revert to old patterns the moment leadership attention moves elsewhere.

Leading change begins with narrative. A clear and compelling explanation of why the organization is evolving, what outcomes are expected, and how individuals will be affected helps to align stakeholders across diverse roles. The most successful narratives are rooted in the impact of their mission rather than technical jargon. For example, instead of describing a new platform as a way to consolidate data lakes, leaders should present it as the means by which an analyst will cut the time needed to produce a briefing from three days to three hours. In commercial organizations like Microsoft, Satya Nadella's early emphasis on a growth mindset and empowerment reframed the company's culture from one of competition to one of collaboration. This was not a slogan but a deliberate repositioning that guided structural and technological changes that followed.

Beyond narrative, change requires mechanisms. Weekly demonstrations of new capabilities, monthly operational reviews, and quarterly portfolio resets give the change effort rhythm and predictability. These rituals offer stakeholders the opportunity to witness progress, express concerns, and gain a deeper understanding of their role in the unfolding transformation. In many leading firms, such as Procter & Gamble, leaders created regular business review cycles that integrated digital transformation metrics directly into business outcomes, ensuring that progress on technology was inseparable from progress on mission.

Operational realities often make change difficult. Existing systems are entrenched, personnel are accustomed to familiar workflows, and external regulators impose requirements that may slow adoption. To address this, successful organizations employ transitional strategies. Dual run windows, where old and new systems operate in parallel, build confidence and reduce risk. Temporary scaffolding, such as connectors that enable legacy systems to integrate with modern platforms, can buy time while minimizing disruption. In government contexts, agencies like the IRS have utilized dual running periods during major modernization efforts to ensure that critical taxpayer services were not interrupted. At the same time, new digital systems were brought online.

Another important dimension is the human element. Change is experienced differently across the workforce. Some see new tools as opportunities, while others perceive them as threats. Effective change leaders identify and cultivate champions who can influence their peers, while also creating safe spaces for skeptics to express concerns and receive support. Programs such as train-the-trainer, office hours, and peer learning networks help employees adapt in ways that feel collaborative rather than imposed. During its shift toward cloud-centric operations, Google invested heavily in internal champions and communities of practice to ensure that new engineering practices were consistently adopted across thousands of teams.

Leadership must also address incentives. Formal performance measures should be aligned with the behaviors required for successful adoption. Recognition should not only reward feature delivery but also emphasize risk reduction, automation of compliance, and demonstrated improvements in decision outcomes. At Capital One, performance metrics for technology leaders were directly tied to both delivery speed and regulatory compliance, ensuring that governance was not treated as an afterthought.

Change leadership further involves managing external stakeholders. Regulators, oversight bodies, and partners often need reassurance that transformation is being conducted responsibly. Transparent communication, automated evidence collection, and proactive reporting help reduce resistance and build trust. In defense settings, this may involve demonstrating that model risk controls are not only theoretical but also operationally embedded. In commercial settings, companies like Salesforce have invested in transparent trust dashboards to communicate security and uptime data to customers, reinforcing confidence during periods of rapid change.

Finally, change leadership must be sustained. Initial enthusiasm can wane, and organizations risk reverting to old habits. Leaders must therefore embed feedback loops that capture adoption data, user satisfaction, and the impact on their mission. These feedback loops inform adjustments, ensuring that the change process evolves in line with emerging needs and technologies. The most effective organizations treat change not as a project with a start and finish but as a core competency. By doing so, they position themselves to continually adapt as the external environment shifts.

Leading change in practice requires vision, structure, empathy, and persistence. It is about telling a story that resonates, establishing rhythms that reinforce momentum, managing the dual demands of continuity and innovation, addressing the diverse experiences of the workforce, aligning incentives, engaging external stakeholders, and maintaining a feedback loop for continuous improvement. When these elements come together, organizations not only implement new systems and policies but also evolve their very capacity for adaptation, ensuring they can meet the next wave of technological and mission requirements with confidence.

Measuring What Matters

One of the most persistent challenges in data and AI transformation is determining what to measure and how to measure it effectively. Metrics can either accelerate progress by focusing energy on outcomes that matter, or they can distract and even cause harm if they reward the wrong behaviors. For a CDAO, building a disciplined approach to measurement is essential to ensure that transformation translates into real mission or business impact.

The first step is clarifying what truly matters. Vanity metrics such as the number of models deployed, terabytes ingested, or dashboards created may look impressive on paper, but often say little about whether decisions have improved. The more meaningful questions are: Are operators and leaders making better decisions more quickly? Has risk been reduced? Has mission throughput improved? These are the outcomes to which measures must ultimately be tied. For example, in a logistics context, the critical measure may be how quickly supply decisions can adjust to disruptions, while in a healthcare setting, it may be the reduction in time from diagnosis to treatment for at-risk patients.

Once priorities are clear, the next challenge is capturing data to support measurement. Too often, organizations neglect to design instrumentation into systems from the beginning, leading to costly retrofits later. By embedding telemetry into workflows, product teams can automatically collect data on usage, time to completion, error rates, and satisfaction. Commercial firms like Netflix and Amazon have long mastered this practice, using continuous instrumentation to understand not just what customers click, but how long they engage and whether those interactions drive retention and revenue. In the federal sector, defense organizations increasingly embed telemetry into command-and-control systems to monitor decision-cycle times, ensuring that transformation efforts are objectively evaluated.

Collecting metrics is not enough; leaders must also communicate them effectively. Numbers without context can alienate or confuse stakeholders. Effective communication ties metrics back to mission outcomes and frames them in stories that resonate. For example, rather than simply reporting that model accuracy improved by three percent, a leader might explain that the improvement allowed analysts to clear twice as many cases in the same amount of time. Dashboards should not overwhelm, but should highlight the few measures that indicate whether change is on track. Quarterly reviews should translate data into decisions about where to invest more and where to adjust course.

Another persistent challenge is data quality. Leaders can become paralyzed by doubts about the reliability of their measures. The solution is not to seek perfect data before acting but to establish methods for validation and triangulation. Sampling, independent audits, and comparison with ground truth when available can provide reasonable assurance. Importantly, the act of measuring should itself improve quality, as teams understand that their outputs are subject to scrutiny. Google, for example, found that once engineering teams knew that deployment success rates were being monitored, they improved testing discipline, which in turn raised both quality and trust in the metrics.

There are also cautionary tales. Organizations that measure the wrong things can create inverse incentives. A well-known case occurred in call centers that measured only call length, leading agents to rush customers off the line without resolving problems. In the AI era, companies that count only the number of models deployed may flood their operations with poorly governed solutions, creating compliance and reputational risks. The Wells Fargo scandal, in which employees created false accounts to meet aggressive sales metrics, illustrates the danger of rewarding outputs without ensuring they truly serve the mission or customer goals. These examples underscore the need to validate not just the accuracy of data but also the alignment of measures with desired outcomes.

When applied responsibly, effective measurement becomes a feedback loop. It allows leaders to evaluate whether structural and cultural changes are delivering value, to adapt approaches that are underperforming, and to reinforce behaviors that lead to success. Strong measurement cultures encourage transparency: both successes and shortcomings are visible, and both are used to improve. The CDAO can foster this by insisting on open reporting, by making metrics accessible to the workforce, and by framing measurement as a tool for learning rather than punishment.

Measuring what matters requires leaders to move beyond vanity metrics and anchor evaluation to mission outcomes. It requires embedding telemetry into workflows from the outset, communicating results in ways that tie back to decisions and impact, validating quality without becoming paralyzed by imperfection, and resisting the temptation to reward volume over value. When organizations do this well, they not only track progress but actively accelerate it, building a culture of accountability and continuous improvement. When they do it poorly, they risk undermining trust, wasting resources, and even causing harm. The choice of which path to follow lies squarely with leadership.

QUICK NUMBERS
The Cost of Measuring the Wrong Things

- A 2022 survey of contact center executives found that over **45%** of organizations that tracked call time as their primary performance metric experienced declines in customer satisfaction, underscoring how misaligned measures can backfire. (Marr 2022[21])
- The Wells Fargo account scandal, driven by aggressive sales targets, led to over **$3 billion** in fines and remediation costs, underscoring the financial and reputational risks of incentivizing the wrong outcomes. (Cowley 2020[22])
- By contrast, Google's Site Reliability Engineering practice demonstrated that aligning measurements with reliability and user experience improved both uptime and developer efficiency, with some teams reporting **up to 40% faster deployment cycles** once error budgets were introduced. (Beyer 2016[23])

These examples highlight the importance of selecting metrics that reinforce mission outcomes rather than create perverse incentives.

From Program to Institution

Many teams can stand up a promising program. Far fewer convert that spark into an institution that survives leadership changes, shifts in strategy, and the relentless churn of new technologies. The difference is not a matter of scale alone. It is the conscious shift from building isolated wins to establishing repeatable mechanisms that produce wins on demand. For a Chief Data and AI Officer, this is where organizational design matures into organizational resilience.

Institutionalization begins with a specific, measurable purpose tied

21. Marr, Bernard. "Call Center Metrics: Choosing the Right KPIs for Customer Experience." Forbes, 29 Mar. 2022, www.forbes.com/sites/bernardmarr/2022/03/29/call-center-metrics-choosing-the-right-kpis-for-customer-experience .
22. Cowley, Stacy. "Wells Fargo to Pay $3 Billion over Fake Accounts." The New York Times, 21 Feb. 2020, www.nytimes.com/2020/02/21/business/wells-fargo-fake-accounts-settlement.html .
23. Beyer, Betsy, et al., editors. Site Reliability Engineering: How Google Runs Production Systems. O'Reilly Media, 2016.

to decisions that matter. New efforts are often tempted to build broad platforms or sprawling roadmaps. The programs that become institutions start narrower. They select a small number of urgent, consequential decision points, then deliver products that change those decisions in visible ways. The point is not only to ship features but to demonstrate a way of working. Telemetry is embedded from the start so that impact can be observed rather than asserted. Policy is translated into code so that safety is automatic rather than manual. The operating rhythm connects delivery to real review cycles, so leaders see progress and can steer. When these mechanisms are visible and repeatable, they begin to create trust. Trust is the raw material that allows programs to graduate into institutions.

Building from scratch requires patience and sequencing. Early wins create permission to standardize. Standardization creates the foundation for scale. At this stage, leaders capture the patterns that worked and turn them into golden paths. Naming conventions, data contracts, model release checklists, and evidence pipelines stop being heroic one-offs and become the default path for every team. This does not slow innovation. It removes friction and cognitive load. In the private sector, companies like Capital One and Microsoft have shown that when platform rules are clear, engineers move faster because the hard parts are handled for them. In the public sector, organizations such as the UK Government Digital Service demonstrated that a small set of common standards could raise quality across dozens of services while shortening delivery cycles. The lesson is consistent. Institutions are built when the organization can say, "This is the way we work, and that way is both safe and fast."

Evolving an older program into an institution requires a different kind of leadership. Legacy systems and contracts often carry high sunk costs and political commitments. The CDAO must separate sentiment from signal. Start with an honest inventory of what exists, including technology, processes, and obligations. Identify the assets that still drive decision outcomes and elevate them. Identify the patterns that cause delays or risks and set deprecation plans with concrete dates and replacements. Deprecation is not a press release. It is a managed process that keeps users whole. Dual running reduces anxiety by allowing the new to prove itself while the old continues to function. Strangler patterns wrap legacy workflows with stable interfaces, allowing migration to proceed in steps rather than in leaps. Across several defense and civil agencies, this approach enabled modern platforms to absorb critical functions without jeopardizing service delivery. In the private sector, firms such as Adobe and Intuit used progressive migration to shift entire product lines to cloud architectures while protecting revenue and the customer experience.

What makes an institution durable is not only artifacts but the way people learn and are rewarded. Training must explain not just tools but doctrine. Teams should understand why the organization designs around decisions, how governance as code protects both speed and safety, and what good looks like in release and operations. Acquisition language must reflect those lessons. Contracts that once purchased isolated systems should now require semantic interoperability, continuous telemetry, and automated evidence as standard deliverables. Performance systems should recognize more than feature throughput. They should reward reduction of operational risk, adherence to golden paths, and measurable impact on mission or business outcomes. When these expectations live in training, contracts, and evaluation, they stop being leadership preferences and become organizational habits.

Institutions endure because they learn. Technology and missions change. Standards that were once helpful can become constraints. A resilient organization creates a cadence for renewal that does not destabilize the whole. Architecture reviews are open and humble. Teams can propose changes to standards with data, and approved changes are versioned and communicated. Portfolios are rebalanced on a regular schedule so capacity flows toward high-value outcomes rather than lingering in pet projects. User feedback is a first-class signal, gathered through champion networks, office hours, and product telemetry. This learning posture keeps the institution aligned with reality rather than with last year's plan.

The CDAO is uniquely positioned to steward this progression from program to institution. The role spans data, analytics, AI, governance, and the platform, allowing the CDAO to see the system as a whole. That vantage point is essential for arbitration. There will be moments when speed conflicts with safety, when local optimization conflicts with enterprise standards, or when an innovation should be allowed to diverge in order to discover a better future pattern. The CDAO's job is to set guardrails, make the tradeoffs explicit, and ensure that evidence drives decisions about what to scale and what to retire. The posture is pragmatic. Keep what works. Fix what can be improved. Retire what holds the mission back.

Institutionalization is also a shift in budgeting and ownership. Programs often rely on exceptional funding and heroic leaders. Institutions rely on durable lines of effort and shared ownership. Outcome-based funding replaces one-time capital projects with tranches tied to demonstrated impact and risk retirement. Showback and chargeback create transparency so consumers of platform services understand costs and can make informed choices. Vendor ecosystems are treated as extensions of the platform rather than as bespoke projects. Open interfaces and shared ontologies reduce lock-in and invite partners to contribute modules that others can reuse. These financial and commercial choices reinforce the technical and cultural patterns that make the institution strong.

That said, there are certainly pitfalls. Some programs standardize too early and freeze practices that are not yet proven. Others never standardize and drown in variation. Some cling to legacy because the migration is uncomfortable. Others chase novelty and leave behind fragile systems. Institutions avoid these traps by keeping the focus on the decision outcomes that matter and by using evidence to judge progress. If a standard helps teams deliver faster with fewer incidents, keep it and teach it. If it slows teams or hides risk, change it. If a legacy system still delivers value safely, wrap it and plan a clear future. If it creates drag and cannot be secured, retire it with discipline and care.

The destination is not a monolith. It is a set of habits that make the organization stronger each cycle. When a new leader arrives, the cadence continues. When a new vendor appears, the interfaces protect the enterprise. When a new technology emerges, the institution can evaluate, pilot, and either adopt or decline without drama. At that point, the early program has become an institution. Decision advantage is no longer the product of exceptional effort. It is the way the organization works.

LOOKING AHEAD
From Programs to Institutions

Many organizations celebrate the launch of a new program but fail to establish the habits that make it last. The transition to an institution requires moving from exceptional funding and ad-hoc champions to repeatable mechanisms, codified standards, and embedded governance.

Examples from both government and industry demonstrate this shift. The UK Government Digital Service transformed one-off digital projects into an institutional model by standardizing design patterns and procurement language across agencies (2012[24]). In the private sector, Microsoft institutionalized its cloud transition by weaving compliance and telemetry into every release cycle, ensuring that practices outlasted leadership changes. (Nadella, Shaw 2017[25])

For the CDAO, the lesson is clear: a program is a promise, but an institution is a practice. The ability to evolve from one to the other is the hallmark of organizational resilience.

Lasting Organizational Resilience

This chapter has traced the journey from the initial challenge of designing organizations around decisions to the deeper responsibility of leading change, measuring impact, and converting programs into institutions. We began with the premise that structure should follow decisions, not data, and we saw, through examples in both the public and private sectors, how aligning teams with decision outcomes creates agility and resilience. We explored the archetypes of organizational form and considered how each might be operationalized, adapted, and blended over time. We then examined closely the roles and responsibilities that define a modern data and AI

24. United Kingdom, Government Digital Service. Government Digital Service Design Principles. gov.uk, 2012, www.gov.uk/guidance/government-design-principles .
25. Nadella, Satya, with Greg Shaw, and Jill Tracie Nichols. Hit Refresh: The Quest to Rediscover Microsoft's Soul and Imagine a Better Future for Everyone. Harper Business, 2017.

leadership ecosystem, focusing on how clarity and alignment prevent overlap and confusion. From there, we turned to the realities of leading change in practice, highlighting the rhythms, narratives, and incentives that sustain momentum. We examined the discipline of measuring what matters, cautioning against vanity metrics and emphasizing feedback loops that foster continuous learning and improvement. Finally, we considered the arc from program to institution, recognizing that resilience emerges when successful practices become embedded habits rather than temporary projects.

Taken together, these threads underscore that organizational design and change leadership are not abstract exercises but operational imperatives. The CDAO must be a strategist, architect, and storyteller, but also a pragmatist who knows when to codify, when to adapt, and when to discard. The measure of success is not whether a single platform or product launches, but whether the organization gains the capacity to continuously deliver decision advantage in the face of evolving missions, technologies, and constraints.

Looking forward, the next chapters of this book build on this foundation. Where this chapter has focused on internal design and the leadership of change, subsequent chapters will extend outward. We will explore how data and AI strategy connect with external ecosystems, including vendors, partners, and regulators. We will turn to questions of ethics, risk, and trust, asking how organizations can ensure responsible use of AI at scale. We will also examine talent, workforce development, and the cultural dimensions that sustain innovation over decades rather than years. Later chapters will revisit the theme of measurement, extending it to portfolio governance, investment strategy, and cross-organizational alignment.

In this sense, the work of organizational design and change leadership is both a destination and a departure point. It is a destination in that it produces resilient structures and habits that endure. It serves as a departure point, as those very structures become the launchpad for broader transformation, enabling the CDAO and their teams to tackle the next frontier of challenges with confidence. Readers should carry forward from this chapter not only practical ideas for structuring teams and leading change, but also the conviction that organizational resilience is possible, that it can be deliberately built, and that it is the foundation on which every other aspect of data and AI leadership will depend.

Chapter 5

DATA GOVERNANCE & POLICY ALIGNMENT

The previous chapter on organizational design made one point abundantly clear: structure shapes outcomes. Decisions about how teams are arranged, how authority is distributed, and how accountability flows will either accelerate or slow the use of data and AI. Yet structure alone is not enough. Even the best-designed organization will stumble if its data is unreliable, inaccessible, or surrounded by confusion about what can and cannot be done. This is where governance and policy alignment enter the story. They are the connective tissue that ensures a well-designed organization can operate at speed and with confidence.

For years, governance carried an unfortunate reputation. To many, it was a "dirty word" that conjured images of endless committees, thick binders of rules, and months of delay before anyone saw value. Mission owners often felt that governance was something done to them rather than with them. Analysts and operators whispered about it in conference rooms, using it as a synonym for "slow." In that era, governance was viewed as a compliance exercise, separate from the actual work of delivering outcomes.

That outdated view is no longer tenable. The demands placed on today's organizations, public and private sector alike, require governance that moves at the same pace as missions and markets. As we saw in earlier chapters, trust in data is the ultimate currency, and cultural alignment is what transforms ambition into daily practice. Governance brings those elements together. It is the mechanism that makes data trustworthy, policy actionable, and decisions defensible. Without it, even the most elegant structures collapse under the weight of inconsistency and doubt.

What is emerging now is a different vision. Agile Data Governance[26] treats governance not as a hindrance, but as an enabler of faster, smarter, and safer decisions. It distributes responsibility rather than hoarding it in a single office. It embeds rules and standards into the very tools people already use, making the compliant path the easiest one. It uses automation and AI not as replacements for human judgment, but as amplifiers of stewardship. In this model, governance is not an obstacle to innovation. It is the quiet infrastructure that makes innovation possible.

The role of the Chief Data and AI Officer is not to reinvent governance from scratch or to drown the enterprise in policies. It is to translate policy into practice, aligning authority with mission, and empowering teams to act with clarity. Done well, governance becomes invisible, like the foundation of a building; rarely noticed, but impossible to live without.

This chapter will explore how to make that vision real. We will examine how policy serves as the north star, how governance has evolved into a distributed practice, how quick wins build credibility, how technology can be used as an enabler rather than a burden, and how organizations can evolve thoughtfully over time. Along the way, we will examine examples where governance accelerated mission success and others where outdated models led to failure. The lesson throughout is simple but profound: governance is not someone else's problem. It is everyone's responsibility, and when embraced as such, it becomes one of the most powerful tools a leader can wield.

26. Loyens, Jon. Agile Data Governance. data.world, 2020, apexassembly.com/wp-content/uploads/2020/11/Agile_Data_Governance_white_paper.pdf .

QUICK NUMBERS
The Cost of Outdated Governance

- **67% of organizations** report that overly centralized governance structures led to a rise in "shadow IT" or parallel data systems. (Gartner, 2022[27])
- **45% of enterprises** admit that governance processes relying heavily on documentation and manual reviews delayed analytics projects by more than three months. (Forrester, 2023[28])
- **60% faster adoption rates** are achieved by organizations that roll out governance incrementally with visible early wins, compared to those attempting enterprise-wide "big bang" launches. (IDC, 2022[29])
- **2 out of 3 failed governance initiatives** cited "one-size-fits-all policy enforcement" as the primary cause of user resistance. (McKinsey, 2021[30])
- Companies that combined automation with distributed stewardship reported **a 40% reduction in compliance costs** while improving audit performance. (Deloitte, 2023[31])

TAKEAWAY: Governance models that rely on central bottlenecks and paper policies breed shadow systems and slowdowns. Adaptive, distributed, and automated approaches not only drive adoption but also reduce costs and risk.

27. Gartner. Data and Analytics Governance Survey 2022. Gartner, 2022.
28. Forrester. The State of Data Governance, 2023. Forrester Research, 2023.
29. International Data Corporation (IDC). Worldwide Data Governance Adoption Trends, 2022. IDC, 2022.
30. McKinsey & Company. Why Data Governance Programs Fail. McKinsey & Company, 2021.
31. Deloitte. Automation and Stewardship: Redefining Data Governance for Compliance and Performance. Deloitte Insights, 2023.

Policy as the North Star

Every transformation described in the earlier chapters rests on one essential foundation: clarity of purpose. We emphasized that leadership provides vision and accountability, and demonstrated how organizational design establishes the framework for decision-making. Neither of those elements can succeed if the enterprise lacks a clear set of rules that defines the boundaries of authority and the expectations of behavior. This is the role of policy.

Firm policy is not only a constraint but also a source of empowerment. Without it, governance challenges devolve into personal interpretation, competing preferences, or inconsistent enforcement. With it, governance has legitimacy. Policy tells leaders what is required, where risk tolerance lies, and how compliance should be demonstrated. Policy alignment ensures that governance is not arbitrary but grounded in widely recognized authority.

Too often in the past, policy has been viewed as merely paperwork rather than as a vital operational tool. Agencies and enterprises issued thick binders of rules that were quickly shelved or translated into rigid checklists. These "paper policies" often created more confusion than clarity. They left employees uncertain about what mattered most and generated bottlenecks without necessarily improving outcomes. This echoes one of the recurring lessons from the "Lessons Learned" section later in this chapter: paper policies are insufficient on their own.

When CDAOs anchor governance in policy, they unlock a powerful accelerator. Policies can simplify choices by clarifying priorities. They can act as cover for leaders who need to drive difficult changes but require visible authority to do so. When the Department of Defense released its Data Strategy, it created a lexicon and guiding principles that helped the Services align their efforts with confidence rather than with hesitation. Civilian agencies have leaned on OMB memoranda and the Federal Data Strategy Action Plans to do the same. These touchpoints enabled leaders to argue not only that a change was smart, but that it was required by higher authority.

Policy also sets the boundaries of risk appetite. A financial regulator will restrict access tightly to protect confidentiality. A research institution may lean toward openness to accelerate collaboration and discovery. Governance practices follow directly from these choices. When risk tolerance is not explicit, organizations fall into one of two traps: they either over-govern and slow innovation, or they under-govern and invite noncompliance.

In practice, policy alignment requires deliberate translation. Few staff members will read an executive order or statutory requirement in full. The CDAO's role is to interpret these policies in plain language and demonstrate how they align with real-world workflows. This translation should also be contextual. In the Federal market, the Evidence Act provides broad guidance on data accessibility, but each agency must interpret how that guidance applies to its mission. HIPAA provides detailed privacy requirements for health data; however, healthcare organizations must still determine how these requirements apply to their clinical and research environments. Those who treat HIPAA as a static rulebook often slow down patient care without necessarily improving security. Those that embed its principles into automated audit trails and access controls create a culture of trust where clinicians and researchers can move with confidence.

Policy also rarely arrives in a single layer. International agreements, national laws, sector-specific regulations, and organizational directives can overlap or conflict. A defense organization may need to reconcile U.S. classification policies with allied information-sharing agreements. A bank may need to comply with both domestic regulations and international privacy regimes, such as the GDPR. Governance aligned with policy must not only enforce a single rule but also reconcile multiple layers in a way that avoids uncertainty for employees.

The most important lesson is that policy cannot be static. Just as our next chapter on technology modernization emphasizes the need for continuous adaptation, so too must governance frameworks evolve in response to changing policies. Technology advances, new threats emerge, and societal expectations shift. Policies that once seemed sufficient may become inadequate. CDAOs must therefore treat policies as living instruments, regularly reviewed and refreshed. Governance practices should not be tied to a rigid rulebook, but rather designed to adapt as the policy environment evolves.

Operationally, CDAO leaders can take several steps to anchor governance in policy:

- **Map policy to practice.** Create explicit linkages between specific regulations, executive orders, or internal directives and the governance processes that support them. This helps employees understand not just what they must do, but why it matters.

- **Use policy to prioritize.** Policies often outline critical mission or compliance obligations. Align governance efforts first to those priorities, rather than trying to boil the ocean. This ensures early wins and visible value.
- **Translate policy into plain language.** Dense policy documents can be inaccessible to most employees. The CDAO's role is to act as a translator, explaining policy requirements in clear, actionable terms that resonate with practitioners.
- **Establish living policies.** Policies should not be static artifacts. In a rapidly evolving data and AI environment, policies must be regularly reviewed and updated to reflect new technologies, emerging threats, and changing opportunities. Governance must adapt in tandem.

Ultimately, policy provides both direction and legitimacy. It is the North Star for governance, guiding organizations away from arbitrary choices and toward mission-aligned authority. By beginning with policy, CDAOs can give shape to their governance efforts, reduce organizational resistance, and create systems that not only comply but also accelerate decision-making. Governance rooted in policy is not an obstacle. It is the foundation that allows leaders to build trust in data, deliver on strategy, and make organizations resilient in the face of change.

QUICK NUMBERS
Why Policy Alignment Matters

- **65% of organizations** report that unclear or conflicting data policies are the single biggest barrier to effective governance. (Gartner, 2023[32])

32. Gartner. Top Data and Analytics Governance Trends 2023. Gartner, 2023.

- **Up to 30% of IT budgets** are wasted on duplicate systems or rework when governance is not anchored in a clear policy direction. (McKinsey, 2022[33])
- **42% faster compliance audits** are achieved in organizations that automate policy enforcement within their governance frameworks. (IDC, 2023[34])
- **3 out of 4 federal CDOs** cited "policy alignment" as the most critical success factor for scaling data strategy across agencies. (Federal CDO Council, 2022[35])
- Agencies that connected governance directly to statutory requirements, such as the Evidence Act or FISMA, reported **twice the adoption rate** of data-sharing initiatives compared to those that treated governance as a standalone function. (Federal Data Strategy Progress Report, 2021[36])

TAKEAWAY: Clear policy alignment is not just a compliance box to check. It is a performance accelerator, reducing wasted effort, speeding audits, and giving leaders confidence that their governance decisions are both defensible and effective.

Governance as an Organizational Practice

In earlier chapters, we explored how strategy provides direction and how organizational design creates the structure through which decisions flow. Governance must be understood as the practice that ties these together. Without governance, strategy becomes aspiration without accountability, and structure becomes form without function. Governance is not a theoretical overlay. It is the lived discipline that ensures decisions are made on trusted data, in alignment with policy, and in service of mission outcomes.

33. McKinsey & Company. The Data-Driven Enterprise of 2025. McKinsey & Company, 2022.
34. International Data Corporation (IDC). Worldwide Data Governance and Compliance Automation Forecast, 2023–2027. IDC, 2023.
35. Federal Chief Data Officers Council. Annual Report 2022. Chief Data Officers Council, Office of Management and Budget, 2022.
36. U.S. Office of Management and Budget. Federal Data Strategy 2021 Progress Report. Executive Office of the President, 2021.

For years, centralization was treated as the remedy for every governance failure. When data was scattered, we centralized it. When teams made inconsistent choices, we centralized authority. Each reform promised coherence, yet each produced new layers of friction. The lesson is not that centralization is wrong; it's that *overcentralization* trades speed for symmetry.

The goal of modern data governance is not control, but coordination. The most effective organizations operate as *federated systems*: decision-making is distributed, but bound by shared standards, common data definitions, and transparent oversight. This balance allows mission teams to act quickly while ensuring that what they produce remains interoperable and accountable at scale.

In this model, central functions still play a vital role. They set the rules of the road, manage the shared infrastructure, and ensure that the organization learns collectively. But the authority to move (the power to decide and to act) resides at the edge, closest to where the mission happens. Centralized visibility, not centralized control, is what makes federation work.

From Centralized Control to Distributed Responsibility

The concept of distributed governance has been advanced most prominently through the idea of **Data Mesh**, which treats data as a product and assigns responsibility for quality and stewardship to the teams closest to it. This approach does not remove policy or oversight. Instead, it reframes them as shared frameworks within which domains manage their own data. By contrast, the **Data Fabric** approach emphasizes centralized orchestration through unified platforms. Each has value, and most organizations benefit from a hybrid approach: centralized platforms that ensure consistency, combined with distributed ownership that provides accountability and context.

For CDAOs, the key lesson is that governance must be both shared and coordinated. Over-centralization leads to bottlenecks. Pure decentralization leads to fragmentation. The challenge is to balance the two by embedding governance within domains while ensuring alignment with enterprise policy and strategy.

Embedded Governance, Not Administrative Overhead

Governance must also shift from an administrative task to an integrated part of the daily workflow. A data scientist should not need to submit manual requests to classify a dataset. A business analyst should not have to spend hours documenting metadata. A software engineer should not be left guessing about whether a data-sharing decision complies with policy. These tasks should be automated, contextual, and seamless within existing tools.

The lesson here mirrors the modernization principle outlined earlier: the best way to drive adoption is to make the proper behavior the easiest one to adopt. When governance is embedded in systems and workflows, compliance becomes an output of work rather than an additional chore.

One federal health agency provides a valuable example. Instead of requiring developers to check policies in a separate document library, the agency embedded data-sharing rules directly into its API layer. The APIs enforced the rules automatically, ensuring compliance without slowing development. This shift transformed governance from an external requirement to an integral part of the system.

Governance as a Cultural Shift

Technology and policy are necessary, but they are not sufficient. Governance only becomes sustainable when it is embedded in culture. Our earlier exploration of culture showed that it is defined not by slogans but by what people do when no one is watching. A governance culture is one in which staff at every level see themselves as stewards of the data they use.

Creating this culture requires intentional actions by leadership:

- **Empower stewards within domains.** Stewardship should be a recognized responsibility, not a part-time burden. Provide time, authority, and recognition so that stewards succeed.
- **Promote shared accountability.** Reinforce that governance is not the responsibility of a single office but of the entire organization. Celebrate examples of strong stewardship by teams.
- **Create transparent feedback loops.** Allow data consumers to flag issues, report errors, or suggest improvements directly to data producers. This keeps governance adaptive and collaborative.

- **Model stewardship from the top.** Executives should ask about data lineage in briefings. Managers should give time to stewards. Analysts should cite their data sources in reports. Each act signals that stewardship matters.

The financial services sector offers a cautionary and hopeful example. Many firms were fined heavily for poor data quality and compliance failures. Those who responded with more rules and heavier documentation often made matters worse. Others took a different approach, embedding data stewards within business units and providing them with both tools and authority. Over time, these firms turned governance into a competitive advantage by delivering more accurate insights, faster audits, and improved customer trust.

Prescriptive Guidance for CDAOs

For a Chief Data and AI Officer, operationalizing governance as a living practice requires both vision and discipline. The following principles provide a practical framework:

- Start with domains where governance pain is most acute, and success will be visible. Build credibility with focused improvements before expanding enterprise-wide.
- Balance centralized platforms that enforce baseline policy with distributed ownership that brings context and accountability.
- Embed governance into existing tools and workflows so that compliance is the natural outcome of doing the job correctly.
- Measure and celebrate progress, not just in compliance metrics but in improved mission outcomes such as faster access, higher trust in data, and reduced rework.
- Translate policy into practice, using clear language and automation so that staff understand both the "what" and the "why" behind governance.

Beyond Compliance Toward Enablement

Ultimately, governance as an organizational practice is about much more than reducing risk. It is about creating trust that enables speed. A well-governed organization can move faster because its people are confident in the data they use. Decisions are better-informed, innovations scale more quickly, and compliance is achieved without slowing the delivery process.

When governance is distributed across domains, embedded in workflows, reinforced through culture, and supported by technology, it ceases to be an obstacle and becomes a catalyst. It is not paperwork. It is practice. It is not someone else's responsibility. It is everyone's. In that sense, governance is not just a requirement but one of the most potent enablers of mission success that a CDAO can champion.

VOICES IN THE FIELD
Shaping Modern Governance

"Data governance cannot be effective if it is seen as a centralized policing function. It needs to be federated, automated, and owned by the teams who know the data best. When governance becomes a natural part of how teams build and share data products, it finally starts to scale."
–Zhamak Dehghani, Creator of Data Mesh

"Agencies that empower mission units to take ownership of their data assets, while providing centralized platforms for policy enforcement, are seeing higher adoption and faster progress on the Federal Data Strategy goals."
–Federal CDO Council, 2022 Annual Report

"The worst thing you can do to an organization is make governance synonymous with bureaucracy. The best thing you can do is make it synonymous with trust. When people trust the data, they move faster, they take smarter risks, and they innovate more."
–DJ Patil, Former U.S. Chief Data Scientist

> "We realized governance could not be managed by a central team. The only way forward was to embed stewards directly into each line of business, give them the right tools, and make governance a part of their job description. Once we did that, data quality stopped being someone else's problem."
> –Chief Data Officer, Fortune 100 Financial Services Firm

Quick Wins Without Disruption

Governance programs often fail because they aim for sweeping transformation from the start. Leaders establish councils, roll out new rules across the enterprise, and create layers of documentation. On paper, this looks impressive. In practice, it overwhelms the organization, slows delivery, and undermines trust before any benefits are visible. Staff quickly conclude that governance is more of a burden than a value.

The smarter path is to begin with small, visible wins that prove governance can remove friction rather than create it. This approach builds on themes from Chapter 3, where cultural resilience was shown to grow from steady reinforcement rather than from dramatic resets. By showing progress in weeks instead of years, leaders earn credibility and generate demand for further change.

Why Quick Wins Matter

Quick wins serve three purposes. First, they validate that governance can make work easier. Second, they build trust with stakeholders who may have grown cynical after past failed initiatives. Third, they create momentum that allows for broader adoption without heavy-handed mandates. When people see results in their daily workflows, they are more likely to become advocates for governance rather than skeptics.

Quick wins should be chosen carefully. They should solve real pain points that are widely felt but relatively simple to address. They should also align with policy to strengthen both compliance and efficiency. Most importantly, they should not require a massive organizational redesign.

The Role of Automation

As explored in the previous section on technology, automation is often the fastest way to deliver early results. In the past, governance was associated with manual checklists and labor-intensive reviews. Modern tools can shift this dynamic by embedding policy directly into workflows. Automation does not remove human judgment, but it eliminates repetitive tasks that previously drained time and attention.

Examples include:

- **Automated metadata capture** that records context as data is created, reducing the need for staff to document every detail manually.
- **Policy-aware access controls** that enforce clearance and role rules automatically, cutting approval times from weeks to hours.
- **AI-assisted classification** that tags sensitive fields upon ingestion, preventing compliance gaps without additional paperwork.
- **Embedded data quality checks** in pipelines that prevent flawed data from entering dashboards or reports, reducing rework and errors.

Each of these interventions provides a visible benefit. Staff get faster access to trusted data. Leaders get confidence that compliance rules are being followed. The entire organization sees proof that governance can speed things up rather than slow them down.

Minimizing Disruption

The point of quick wins is not only to solve problems but to do so without interrupting ongoing work. That means embedding governance into tools people already use, rather than creating new portals or processes. It means targeting areas where frustration is high and value is clear, rather than attempting to boil the ocean. Above all, it means respecting the organization's momentum and avoiding heavy reforms that paralyze delivery.

The Census Bureau's modernization of data operations before the 2020 Census offers a strong example. Instead of forcing wholesale change, the Bureau introduced automated disclosure avoidance, metadata capture, and pilot programs in priority domains.[37] Staff experienced immediate relief from manual reviews and siloed approvals. These small, technology-enabled steps generated confidence and created demand for broader adoption. Governance shifted from burden to productivity booster, and leaders gained the trust required to expand further.[38]

Turning Early Success Into Broader Change

Quick wins are not the end goal. They are the footholds that allow the organization to climb higher. Once visible benefits have been delivered, leaders should deliberately expand the scope of their efforts. Each win must prepare the ground for the next, building a sequence of credible improvements.

To turn quick wins into lasting change, CDAOs should:

- **Target problems that matter most.** Choose use cases where staff feel daily pain and where leaders can measure visible outcomes.
- **Align with policy.** Ensure that early wins reinforce existing requirements, thereby building both compliance and trust.
- **Engage users directly.** Design and test solutions with staff to create champions who will promote adoption.
- **Measure impact.** Track improvements in access times, error reduction, or audit readiness to clearly show progress.
- **Communicate broadly.** Share success stories across the enterprise to build momentum and encourage other domains to participate.

The Larger Lesson

Quick wins reduce skepticism, ease the cultural transition, and demonstrate that governance is not just about risk avoidance but about enabling faster, better outcomes. They also create a narrative that supports broader change. Staff who see value in their own work will invite governance into their workflows rather than resist it. Leaders who can point to tangible results

37. Abowd, John M., et al. "The Modernization of Statistical Disclosure Limitation at the U.S. Census Bureau." *U.S. Census Bureau Working Papers*, 2020, www.census.gov/library/working-papers/2020/adrm/modernization-statistical-disclosure-limitation.html.

38. United States, Government Accountability Office. *2020 Census: Innovations Helped with Implementation, but Bureau Can Do More to Realize Future Benefits.* GAO-21-478, June 2021, www.gao.gov/products/gao-21-478.

will have the credibility to expand governance into more complex domains.

Every organization will eventually need comprehensive governance. The challenge is how to get there. Quick wins provide the path. They demonstrate that governance can be cooperative, aligned with policy, and embedded in daily practice. They ensure that change is experienced not as disruption but as progress. And most importantly, they prepare the organization to embrace the broader evolution that follows.

CASE IN POINT
Census Bureau's Bite-Sized Governance Wins[39]

The U.S. Census Bureau faced a daunting challenge: modernize data operations to support the 2020 Census and beyond, while maintaining the highest standards of privacy, accuracy, and compliance. Traditional governance methods (manual approvals, siloed systems, and lengthy review cycles) were far too slow for the scale and speed required.

Instead of attempting a sweeping overhaul, the Bureau rolled out a series of **bite-sized governance wins** that demonstrated immediate value:

- **Automated Disclosure Avoidance:** AI-driven algorithms were deployed to detect and mitigate privacy risks in published data products. This reduced reliance on lengthy manual reviews while enhancing compliance with statutory protections.
- **Metadata Automation:** The Bureau introduced tools to automatically capture and catalog metadata as datasets were created. This made it easier for analysts to discover trusted data sources without asking for centralized approval.
- **Incremental Pilots:** Rather than mandating governance changes across the enterprise, pilots were launched in high-priority domains such as economic indicators. Once those pilots proved successful, other domains asked to adopt similar practices.

39. U.S. Census Bureau. 2020 Census Data Stewardship and Privacy Protections. U.S. Department of Commerce, 2020, www.census.gov/about/policies/privacy/statistical_safeguards/disclosure-avoidance-2020-census.html .

These small, technology-enabled wins changed the conversation. Staff who once saw governance as an obstacle began to see it as a productivity boost. Automated safeguards gave leadership confidence that privacy and compliance were protected without slowing down operations.

By demonstrating value in these early, low-disruption steps, the Census Bureau established the trust and momentum necessary to expand governance more broadly. Automation was not simply a compliance tool; it was the bridge between policy requirements and day-to-day mission execution.

LESSON: Quick, automation-powered wins create buy-in by showing that governance can make work faster, safer, and easier. Once stakeholders experience that value, they become champions for scaling the change.

Technology as an Enabler, Not a Burden

In the next chapter on technology modernization, we'll explore how technology is most powerful when it is paired with process and culture. The same is true in governance. Tools alone do not create trust, compliance, or accountability. At the same time, ignoring technology and relying only on manual oversight makes governance too slow to support missions that demand agility. The role of the Chief Data and AI Officer is to ensure that technology is used to amplify good governance practices without overwhelming staff or creating new silos.

Too many organizations have fallen into the trap of believing that buying a catalog or deploying a lineage tool will solve governance once and for all. The result is often a collection of portals that employees avoid, with adoption lagging and frustration rising. This is why technology should never be treated as the answer on its own. Instead, it must be seen as an enabler that supports policy alignment, cultural stewardship, and distributed responsibility.

Automation as the First Lever

Automation is the most direct way to reduce friction and deliver immediate value. In earlier sections, we saw how quick wins can reshape perceptions of governance. Automation is central to those wins because it eliminates repetitive tasks that previously required human effort. Metadata can be captured automatically as data is created. Access rights can be applied dynamically based on role and clearance level. Data pipelines can prevent flawed records from corrupting dashboards. Automation makes compliance the natural outcome of the work itself, rather than an additional step.

Practical actions include:

- **Automated metadata capture** to reduce manual documentation and make data assets easier to find.
- **Policy-aware access controls** to grant permissions in minutes while maintaining strict compliance.
- **Automated data quality validation** to stop insufficient data from reaching reports, improving trust and reducing rework.
- **Automated disclosure safeguards** to enforce privacy protections at the point of release, speeding delivery of public products while reducing risk.

When deployed in this manner, automation demonstrates that governance can increase both speed and accuracy simultaneously.

Artificial Intelligence as an Amplifier

Artificial intelligence adds another layer of support by extending the reach of human stewards. AI can scan new datasets and classify them as sensitive or public. It can analyze access logs to flag unusual behavior. It can use natural language processing to identify potential policy violations in unstructured documents. These capabilities free people from repetitive scanning, allowing them to focus on judgment and decision-making.

The most effective use of AI is not to replace governance roles but to recommend and guide. AI can suggest classifications, detect anomalies, or highlight risks. Human stewards remain accountable but can respond more quickly and more effectively with machine support.

Knowledge Graphs and Contextual Governance

One of the most challenging tasks in governance is connecting a single dataset to its broader context. A knowledge graph can model relationships between data, policies, people, and processes. It can reveal who owns a dataset, which laws apply to it, how it has been transformed, and where it is being utilized downstream. This context is what enables confident decisions.

For example, a mission planner reviewing logistics data can see immediately that it originates from an authoritative source, falls under a specific policy, and supports multiple dashboards. The planner does not need to guess whether the data is trustworthy. The knowledge graph makes the context visible and actionable.

Other Emerging Tools

Beyond automation, AI, and knowledge graphs, several categories of tools are beginning to shape governance practice. These include:

- **Data fabric platforms** that provide centralized orchestration while allowing distributed domains to enforce their own rules.
- **Automated lineage systems** that map how data moves through pipelines, allowing faster audits and stronger accountability.
- **Self-service governance portals** that let users request access, flag issues, or certify datasets without waiting for a central office.
- **Zero-trust architectures** that extend cybersecurity principles into governance by granting access dynamically based on identity and context rather than static permissions.

Each of these tools can be helpful, but only when deployed as part of a coherent strategy.

Avoiding the Technology Trap

Technology is only as effective as the culture and policy that surround it. If the rules are vague, if accountability is unclear, or if culture resists stewardship, tools will fail. The most common trap is treating governance as a procurement problem rather than as a leadership challenge.

To avoid this trap, CDAOs should:

- Begin with policy alignment and cultural stewardship before adding tools.
- Choose technologies that integrate with existing workflows rather than creating new portals.
- Use technology to solve specific pain points identified by staff, such as access delays or duplicate datasets.
- Expand capabilities gradually, building on successes rather than deploying everything at once.

From Burden to Invisible Infrastructure

When deployed thoughtfully, technology becomes invisible infrastructure. Staff no longer experience governance as an interruption. Access requests resolve automatically because policy-aware controls are embedded in the system. Quality problems appear less often because pipelines catch them in advance. Classification is handled automatically with human review reserved for exceptions.

This is the vision of governance as an enabler. Technology is not the driver but the amplifier. It works in conjunction with policy, process, and culture to create a system that is both trusted and efficient. The role of the CDAO is to orchestrate these elements so that governance becomes not another tool to log into but the very air the organization breathes.

TREND SNAPSHOT
Governance Goes High-Tech

The role of technology in governance has accelerated rapidly in recent years, shifting from an optional add-on to a core enabler of compliance, trust, and speed.

- **70% of large organizations** report they are actively investing in automation to reduce manual governance workloads. (Gartner, 2023[40])
- **56% of agencies** say automated metadata capture has already shortened the time it takes to locate and trust a dataset. (Federal Data Strategy Progress Report, 2022[41])
- **Nearly 50% of enterprises** now use AI tools for at least one governance function, most commonly for sensitive data classification and anomaly detection. (IDC, 2023[42])
- **Organizations that adopt automated lineage tracking** reduce audit preparation time by as much as **40%**, freeing resources for mission- or business-critical priorities. (McKinsey, 2022[43])
- Early adopters of knowledge graph–based governance frameworks report a **30% reduction in duplicate datasets** and a noticeable increase in cross-domain data reuse. (Forrester, 2023[44])

TAKEAWAY: Governance technologies are no longer experiments. They are delivering measurable gains in speed, compliance, and trust. Organizations that fail to harness them risk falling behind not only in innovation but in the basic ability to govern effectively.

40. Gartner. Top Data and Analytics Governance Trends 2023. Gartner, 2023.
41. U.S. Office of Management and Budget. Federal Data Strategy 2022 Progress Report. Executive Office of the President, 2022.
42. International Data Corporation (IDC). Worldwide Data Governance and Compliance Automation Forecast, 2023–2027. IDC, 2023.
43. McKinsey & Company. The Data-Driven Enterprise of 2025. McKinsey & Company, 2022.
44. Forrester. Knowledge Graphs for Data Governance: Emerging Best Practices. Forrester Research, 2023.

Toward Everyday Governance

Therefore, the destination of every governance journey is not another office or thicker binders. It is what we call *everyday governance*: a way of working where trust in data is natural, compliance is quiet, and stewardship is part of everyone's job.

Many organizations have learned this lesson the hard way. Over-centralization creates bottlenecks and shadows. Paper-heavy policies become shelfware. Uniform models ignore fundamental differences among domains and missions. The remedy is not another layer of policy but the steady integration of governance into daily work.

As earlier chapters on leadership and design showed, trust and distributed responsibility are the bedrock of modern organizations. Governance is where these converge. Without it, even sophisticated systems collapse under doubt; with it, organizations gain both speed and confidence.

Building toward everyday governance is less about new tools than new habits. It means aligning policy with daily tasks, embedding automation so compliance happens by design, and nurturing a culture where every domain has a steward accountable to peers. Progress is measured not only by compliance rates but by improved mission outcomes; faster access, fewer reconciliations, more credible analysis.

When governance reaches this state, it becomes the quiet force that sustains resilience and accelerates innovation. People no longer think of governance as bureaucracy. They experience it as clarity; the assurance that the information they use can be trusted.

LOOKING AHEAD
The Next Chapter of Governance Evolution

The next era of governance will not arrive through a single technology or directive. It will come through steady integration of cultural, technical, and policy advances that allow organizations to adapt without losing momentum. Several trajectories are already visible:

- **Self-Governing Systems:** Emerging platforms will increasingly enforce policies automatically in real time, adjusting access, tagging, and lineage without manual intervention. Humans will remain accountable, but much of the heavy lifting will be invisible.
- **Adaptive Policy Engines:** Instead of static rules that require lengthy updates, policy frameworks will become more dynamic, capable of interpreting changes in law or mission requirements and adjusting enforcement accordingly.
- **Human and Machine Collaboration:** AI will take on more of the repetitive monitoring and classification, while humans provide judgment, context, and exceptions. Governance will function more like a partnership than a handoff.
- **Resilient Cultures of Stewardship:** As organizations evolve, the strongest differentiator will not be the sophistication of their tools but the resilience of their culture. Those that normalize stewardship in everyday practice will adapt more smoothly than those that rely on top-down enforcement.

TAKEAWAY: The organizations that thrive will be those that treat governance as a living system. Instead of burning down the old and rebuilding from scratch, they will grow stronger through iteration, weaving in new technologies and practices without sacrificing continuity of mission.

Chapter 6

TECHNOLOGY MODERNIZATION & ARCHITECTURE

The Pulse of Modernization

Technology modernization is not a one-time project. It is a living, breathing discipline that requires constant adaptation to new realities. Technology leadership, such as the CDAO, cannot treat modernization as the procurement of a fixed set of tools; instead, it must view it as building and curating an evolving ecosystem. This ecosystem must be agile enough to support today's mission demands while being resilient enough to grow into tomorrow's challenges.

At its heart, modernization is about relevance. A system that was state-of-the-art ten years ago may today be a bottleneck. A tool that once enabled collaboration may now create silos. An architecture once designed to "do it all" may now do nothing particularly well. Far too many organizations are burdened by outdated infrastructures, sprawling legacy systems, and brittle integrations that struggle to keep pace with mission needs. The pain of these failures is felt in every delayed decision, every duplicated effort, every opportunity lost because the technology did not rise to meet the moment.

This reality makes modernization not optional, but essential. Yet modernization is not only about correcting failures. It is also about unlocking possibilities. Every new generation of technology offers leaders the chance to reimagine workflows, enable faster and more accurate decisions, and connect people and systems in ways that were previously unimaginable. Done well, modernization is not just an upgrade; it is a transformation of how the mission itself can be executed.

And yet, too often modernization efforts stumble. Leaders buy into promises of one-size-fits-all platforms that never deliver. Organizations spend years planning instead of weeks piloting. Users are forced to adopt tools designed around vendor marketing slides rather than their own operational reality. These missteps erode trust, waste resources, and leave the workforce skeptical of the very change they need. The lesson is clear: **modernization done poorly can be worse than no modernization at all.**

The purpose of this chapter is to explore how modernization can be done right; how leaders can approach the challenge with a framework that emphasizes adaptability, user engagement, and mission alignment. It will consider modernization not just as a technical project, but as an ongoing practice that integrates architecture, procurement, and human adoption into a coherent whole. We will explore the challenge of technical debt as a strategic variable that must be actively managed, the role of data architecture as the nervous system that connects every part of the enterprise, and the balancing act between cloud and edge computing that defines so many modern missions.

Modernization is not about chasing trends or adopting technology for its own sake. It is about shaping a living ecosystem that constantly evolves in sync with the mission. It is about building architectures that can flex, adapt, and grow rather than lock the organization into brittle futures. And most importantly, it is about capturing the hearts and minds of the people who will use these systems every day.

The pulse of modernization beats across every organization, whether acknowledged or ignored. For some, it is a steady rhythm that drives continuous improvement and adaptation. For others, it is a faint echo, drowned out by the weight of legacy systems and outdated practices. For leaders ready to listen and to act, the pulse of modernization is an invitation: to reexamine, to rebuild, and to realize the full potential of technology in the service of mission.

A Framework for Modernization

Modernization succeeds when guided by a few enduring principles. This book is built on four foundations: **mission-centered design**, **adaptive architecture**, **rapid procurement and deployment**, and **human-centric adoption**.

These are not rigid steps but interlocking disciplines. Mission-centered design keeps efforts tied to outcomes. Adaptive architecture builds in flexibility. Rapid procurement and deployment compress the distance between need and delivery. Human-centric adoption ensures new tools take root in practice rather than policy.

Taken together, these principles form the scaffolding of modernization. They provide enough structure to maintain coherence while leaving room for adaptation to each organization's mission, culture, and maturity. The rest of this chapter will show how they come to life in practice, drawing lessons from organizations that have turned these ideas into operating reality.

Mission-Centered Design

Modernization often fails when it begins with tools rather than outcomes. Vendors may offer polished solutions that promise to solve every problem, yet the question leaders must always ask is: how does this technology advance the mission? Mission-centered design begins with clarity of purpose. For the Department of Defense, this may mean faster targeting or resilient logistics. For a healthcare organization, it may mean safer patient care or more efficient data sharing across facilities. For the private sector, it may mean delivering a better customer experience or improving operational efficiency.

Expert voices have long argued that technology leadership is only meaningful when tied directly to outcomes. Thomas Davenport and George Westerman's work on digital transformation[45] emphasizes that organizations must embed their modernization efforts into the very fabric of strategy and operations. Similarly, public sector efforts such as the

45. Davenport, Thomas H., and George Westerman. "Why So Many High-Profile Digital Transformations Fail." MIT Sloan Management Review, 9 Mar. 2018, https://sloanreview.mit.edu/article/why-so-many-high-profile-digital-transformations-fail/ .

Federal Data Strategy[46] stress that technology must serve clear, measurable goals rather than become an end in itself.

Adaptive Architecture

No single architecture will serve every need forever. The history of technology is filled with examples of organizations locking themselves into inflexible systems that quickly became obsolete. The lesson from both government and industry is that adaptability must be designed into the system from the beginning.

Adaptive architectures are modular, composable, and built on open standards. They allow components to be swapped out or upgraded without collapsing the entire system. The movement toward cloud-native architectures and microservices in the private sector demonstrates the value of adaptability. Netflix, for example, famously migrated from a monolithic infrastructure to a microservices model, enabling faster innovation and greater resilience. Within the Department of Defense, efforts such as Joint All-Domain Command and Control are similarly focused on creating interoperable, flexible systems that enable information to flow seamlessly across domains and partners (to varying degrees of success, thus far).

Rapid Procurement and Deployment

Traditional acquisition models often leave modernization efforts years behind the pace of innovation. Requirements written at the beginning of a program are outdated by the time a solution is delivered. This mismatch between acquisition cycles and technological change has undermined many otherwise well-intentioned programs.

Modernization requires rapid procurement methods that bring technology into play quickly, test it in realistic settings, and scale only once it has proven its value. Pilot programs, Other Transaction Authorities, and continuous vendor engagement are examples of how government and industry alike are learning to shorten the distance between need and delivery. As agile methodologies have shown in the software world, iterative delivery produces not only faster results but also more relevant solutions.

46. United States, Office of Management and Budget. Federal Data Strategy: Leveraging Data as a Strategic Asset. Executive Office of the President, 2020, https://strategy.data.gov .

Human-Centric Adoption

Perhaps the most overlooked element of modernization is the human one. Too often, organizations assume that if the technology is sound, adoption will follow naturally. Reality proves otherwise. If systems do not fit into workflows, create more friction than they remove, or erode trust through poor performance, users will resist them.

Human-centric adoption begins with involving users early in the process. It requires co-design, feedback loops, and clear communication about why the change matters. It also requires acknowledging the emotional dimension of change. People must see not only how the technology helps the mission, but also how it makes their own work easier, more effective, or more meaningful. Public and private sector experiences alike confirm that user champions, peer advocacy, and visible benefits are more potent than any executive directive.

Putting the Framework Together

When combined, these elements create a flexible framework that adapts to any context: mission-centered design ensures relevance, adaptive architecture ensures resilience, rapid procurement ensures timeliness, and human-centric adoption ensures sustainability.

This framework also guards against the pitfalls that have undermined past modernization efforts. It prevents over-centralization by insisting on modularity. It avoids wasted investment by testing quickly and scaling only what works. It addresses user resistance by engaging people from the start. Most importantly, it provides leaders with a disciplined yet adaptable approach that can withstand the uncertainties of both technology and mission.

The framework's success ultimately depends on what was emphasized previously: trusted data. Mission-centered design provides purpose, adaptive architecture flexibility, and human-centric adoption sustainability, but governance makes progress durable, ensuring that innovation compounds rather than decays.

Technical Debt as a Strategic Variable

Every organization carries technical debt. It may be an outdated system that has not been replaced, a fragile integration that everyone knows might break at any moment, or a patchwork of workarounds that have grown over time. Technical debt is the residue of past decisions. It accumulates when deadlines, mission urgency, or resource constraints force leaders to take shortcuts. It lingers when technologies age out of relevance but remain embedded in workflows. It expands when systems are scaled far beyond their original design.

The language of debt is functional because it emphasizes that these choices are not inherently wrong. Just as financial debt allows an organization to take advantage of opportunities by borrowing against the future, technical debt enables it to move quickly in the present by deferring costs. Shipping a product or capability with a quick solution rather than the perfect one may be the right decision if speed is critical. What matters is not whether debt exists, but whether it is visible, intentional, and managed.

Why Technical Debt Happens

Technical debt often results from three recurring pressures.

1. **Mission urgency**: When speed is paramount, teams may implement temporary fixes in order to deliver on time. In a military context, for example, deploying a minimally viable system to theater may outweigh the risk of future inefficiency.
2. **Resource constraints**: Budget and staffing realities often force compromises. Leaders may choose a "good enough" system today while planning to improve it later.
3. **Evolving technology and requirements**: Even well-designed systems accrue debt over time because the environment changes. Standards shift, new tools emerge, and mission requirements grow. A system that once felt modern eventually becomes a constraint.

Accepting these pressures does not mean surrendering to chaos. It means recognizing that technical debt is unavoidable and managing it strategically.

Managing Debt as a Portfolio

The most effective organizations treat technical debt as a managed portfolio, not an invisible burden. This requires tracking and measuring debt in ways that leadership can understand. Common practices include:

- **Debt registers**: Cataloging known debt items, such as unsupported software, fragile integrations, or missing documentation, and assigning them clear owners.
- **Risk scoring**: Assessing the potential mission impact of each debt item if it fails.
- **Cost visibility**: Quantifying the maintenance costs of debt, such as staff hours spent on workarounds or delays caused by outdated infrastructure.
- **Trend monitoring**: Tracking whether debt is being paid down, accumulating, or shifting into riskier areas.

These practices transform debt from an invisible liability into a visible set of tradeoffs that can inform planning, staffing, and budget decisions.

Strategic Use of Debt

Not all debt is harmful. In fact, some debt is a sign of smart decision-making. Rapid innovation almost always involves taking shortcuts. Startups are famous for building quick solutions that would never scale but are sufficient to test an idea. Government pilot programs often deploy partial solutions to gather feedback before committing to larger investments. In both cases, technical debt is tolerated because the goal is learning and speed.

However, debt becomes dangerous when it is invisible, unmanaged, or misaligned with mission priorities. Systems that quietly accrue complexity may suddenly collapse under operational stress. Workarounds that were meant to be temporary may become permanent, consuming staff time and eroding trust. Debt that once accelerated progress becomes an anchor.

Deciding How to Address Debt

Leaders must decide not only whether to pay down technical debt, but also how to approach it. There are three common pathways:

1. **Aggressively pay down**: Some debt must be eliminated quickly. Examples include unsupported software with security vulnerabilities or systems that pose direct risks to mission success. In these cases, waiting is more costly than acting.
2. **Refactor or write off**: Some debt can be addressed by modernizing components rather than replacing entire systems. Others may simply be decommissioned if they no longer serve the mission.
3. **Allow to accrue intentionally**: Some debt is acceptable if it buys speed or experimentation. For instance, a prototype built with limited scalability may remain in service while its successor is being designed. The key is to acknowledge this debt and plan for its eventual resolution.

Debt as a Planning Indicator

Technical debt should also be viewed as an indicator for modernization planning. High levels of unmanaged debt often signal the need for architectural reform or increased staffing. Tracking the ratio of effort spent on maintaining legacy systems versus building new capabilities can reveal whether an organization is investing in its future or merely surviving its past. Similarly, visible debt portfolios can strengthen budget requests, as leaders can point to measurable risks and inefficiencies that require resourcing to resolve.

The Opportunity of Debt Awareness

By treating technical debt as a strategic variable, leaders shift the conversation from blame to opportunity. Debt is not simply a sign of poor past decisions; it is a natural byproduct of operating in complex, resource-constrained, and rapidly evolving environments. The question is not whether debt should exist, but how it will be managed to enable continued evolution.

Organizations that adopt this mindset find themselves more resilient. They can innovate quickly when needed, knowing that debt is visible and accounted for. They can prioritize paydown when risks grow too high. And most importantly, they can link modernization efforts directly to mission outcomes, ensuring that every dollar and hour spent on reducing debt translates into greater agility and effectiveness.

CASE IN POINT
When Technical Debt Becomes Mission Risk[47]

In 2017, the U.S. Air Force confronted the reality that its Air Operations Center (AOC) 10.1 system had accumulated so much technical debt that it was no longer sustainable. Originally designed years earlier, the system had been patched and extended repeatedly to meet evolving needs. Over time, these fixes created a brittle, complex environment that was expensive to maintain and difficult to adapt.

Attempts to modernize the system under traditional acquisition models only deepened the problem. By the time requirements were finalized, the technology had already fallen behind. A 2018 GAO report noted that the program had spent nearly half a billion dollars without delivering a usable replacement.

The Air Force eventually shifted its strategy, moving toward more agile, modular solutions such as the Kessel Run software initiative. By recognizing technical debt as an existential risk rather than a background nuisance, leaders reframed modernization as a necessity, not a luxury. The lesson is clear: unmanaged debt does not merely slow progress, it can threaten the mission itself.

47. United States Government Accountability Office. Air Force Air Operations Center: Efforts to Transform Program Facing Challenges. GAO-18-130, Government Accountability Office, Jan. 2018, https://www.gao.gov/products/gao-18-130 .

Data Architecture as the Nervous System

If modernization is the body, data architecture is the nervous system. It carries signals, coordinates movement, and allows the enterprise to sense, learn, and respond. Where applications come and go, data architecture endures as the primary interoperability layer, connecting disparate systems, sources, people, and policies. It is the connective tissue that lets a logistics system speak to finance, that lets a sensor feed an analytic model, and that lets a policy become executable logic rather than a PDF on a shelf.

Interoperability First, Tools Second

There is almost never a single system that serves every function of a business. Mergers, mission shifts, and specialized workloads guarantee a heterogeneous landscape. The central design question is not which platform will do everything. The question is how the organization will make many platforms work together. A strong data architecture treats interoperability as a first principle. It favors open standards, clear contracts, and durable interfaces over point-to-point custom wiring. It emphasizes APIs, events, and shared semantics so that teams can add or replace components without breaking the whole.

From Tables to Graphs: Modeling Reality as It Is

Relational schemas are excellent for transactions and well-structured processes. Modern organizations, however, operate across dynamic networks of people, systems, assets, places, and policies. Relationships carry as much meaning as the entities themselves. This is where graphs shine. Graph data models represent entities as nodes and relationships as edges, which maps more closely to how the human brain naturally understands the world.

A knowledge graph elevates this further by layering meaning on top of data. It links datasets through shared concepts, defines relationships with a controlled vocabulary, and encodes business rules as part of the structure. The effect is a living map of the enterprise: what exists, how it relates, and which rules govern its use. When a new system arrives, you connect it to the map rather than rebuilding the map around it.

This vision of knowledge graphs as a unifying layer has deep roots in the history of the web itself. In 2001, Tim Berners-Lee, James Hendler, and Ora Lassila published "The Semantic Web" in Scientific American, articulating a future where data on the web would be machine-readable, not just human-readable.[48] Berners-Lee envisioned a web where information had explicit meaning, where agents could understand relationships between concepts, and where data from disparate sources could be integrated automatically without requiring custom code for every connection.

The Semantic Web vision never fully materialized in the public internet, but its core principles have proven remarkably relevant for enterprise data architecture. Berners-Lee's key insight was that interoperability requires shared semantics, not just shared protocols. HTTP and HTML enabled documents to link, but the Semantic Web proposed going further: enabling data to link through explicit, machine-readable relationships expressed in standards like RDF (Resource Description Framework) and OWL (Web Ontology Language).

What Berners-Lee described for the global web, modern knowledge graphs deliver for the enterprise. Instead of every application maintaining its own isolated understanding of what a customer, order, or asset means, the knowledge graph provides a shared vocabulary and explicit relationships. Instead of writing custom integration code whenever two systems need to exchange data, teams can connect them via the graph's semantic layer. The web standards that emerged from Berners-Lee's vision (URIs to identify concepts, RDF to express relationships, SPARQL to query linked data) have become foundational technologies in enterprise knowledge graphs, proving that his architecture wasn't wrong; it was simply ahead of its time.

The lesson for data and AI leaders is that interoperability at scale requires more than APIs and data lakes. It requires the semantic infrastructure that Berners-Lee recognized as essential: explicit definitions, typed relationships, and machine-readable meaning. Organizations building knowledge graphs today are, in effect, implementing the Semantic Web vision within their enterprise boundaries, creating the interoperable, intelligent data architecture that Berners-Lee saw as necessary for complex systems to work together.

48. Berners-Lee, Tim, James Hendler, and Ora Lassila. "The Semantic Web." Scientific American, vol. 284, no. 5, 2001, pp. 34-43.

Graph Thinking Matches How Humans and AI Reason

Human cognition relies on associations. We remember people in the context of teams, projects, and places. We connect events through cause and effect, not only through time. Modern AI systems, particularly those that retrieve and reason over context, also benefit from this structure. A knowledge graph can serve as an externalized memory that grounds AI with authoritative context. When a model retrieves a policy, a definition, or a lineage path from the graph, it is less likely to hallucinate and more likely to produce relevant, traceable results.

Graph thinking also makes ambiguity safer to handle. If two systems define "customer" differently, the graph can store both definitions, declare where they apply, and show how they relate. Instead of forcing premature uniformity, the graph preserves nuance while still enabling integration.

Semantics as the True Interface

Most integration efforts fail not at the transport layer but at the meaning layer. Two systems can exchange JSON yet still disagree on what the fields represent. A durable data architecture invests in a semantic layer: shared vocabularies, taxonomies, and ontologies that give names and relationships to core concepts. This layer becomes the real interface. It is the promise that "order," "unit," "mission," or "patient" means the same thing wherever it flows, or that differences are explicit and mapped.

When semantics are formalized, governance becomes operational. Policies can be attached to concepts rather than buried in documents. "Only clinicians may access protected health information" becomes a rule bound to a class of data with a defined purpose and lineage. Enforcement then travels with the data through services, pipelines, and queries.

Policies as Data, Not Documents

Treat policies as machine-readable objects tied to the graph. Access controls, retention schedules, data quality thresholds, and model usage constraints can be expressed consistently and evaluated automatically. As datasets move across systems, the architecture evaluates the applicable rules in context. This approach reduces drift between written policy and lived reality. It also simplifies audits because the evidence of compliance is encoded in the same layer that routes and transforms the data.

Data Fabric and Data Mesh: Patterns That Meet in the Middle

Two patterns often surface in modernization conversations. A data fabric focuses on a connective layer that automates integration, metadata capture, and policy enforcement across sources. A data mesh focuses on federated domain ownership with shared standards. These ideas are not in conflict. A practical architecture often blends them. Use mesh principles to place stewardship close to the data and mission, and use fabric capabilities to provide common services such as discovery, lineage, quality checks, and policy application. The knowledge graph sits at the center, providing a shared language for both approaches.

Designing for AI and Analytics from the Start

Analytics and AI are not bolt-ons. They depend on discoverable data, trustworthy lineage, consistent semantics, and timely access. A graph-centric architecture accelerates all of this. It shortens the path from question to answer because relationships are explicit. It improves model governance because inputs, assumptions, and outputs are traceable across domains. It strengthens retrieval for language models because context can be fetched by concept, relationship, or policy, not only by string match.

Practical Interoperability Building Blocks

A nervous system needs standard reflexes. Successful programs tend to invest in a short list of shared capabilities:

- **Canonical entities and reference data**: agreed definitions and codes that anchor integration.
- **Metadata services**: automated capture of lineage, provenance, and quality signals.
- **Event streams**: publish and subscribe patterns that decouple producers and consumers.
- **API gateways and query federation**: consistent access patterns across polyglot stores.

- **Policy decision points**: centralized evaluation of rules that execute close to the data.
- **Identity and attribute-based access control**: authorization that understands roles, purposes, and data sensitivity.

Each capability is small on its own. Together, they allow teams to ship features without reinventing plumbing.

A Short Vignette: Linking the Enterprise through a Graph[49]

Researchers exploring aircraft maintenance and logistics have demonstrated how knowledge graphs can unify complex systems that previously operated in isolation. In one case, data from maintenance records, spare parts inventories, flight schedules, and operational constraints were linked together in a graph-based representation. Instead of forcing a wholesale system replacement, the knowledge graph created a shared layer of context that connected parts, tasks, and operational rules.

With this structure in place, planners could query the graph to answer practical mission questions: which aircraft were grounded due to missing parts, which suppliers offered viable alternatives, and how specific maintenance tasks affected flight readiness. Policies such as export control and mission prioritization were also encoded directly into the graph, ensuring compliance as part of the analytic process.

The result was a more explainable and responsive decision support system. Queries that once required manual cross-checking across multiple databases can now be answered directly, with the graph providing transparent paths showing how each answer was derived. The study highlights how graph-based architectures enable not only technical integration but also faster, more trustworthy decisions in mission-critical environments.

Governance that Moves at the Speed of Change

Strong governance does not slow the nervous system. It keeps it safe. With graph-backed semantics, governance becomes part of the runtime.

49. Olsson, Ella, et al. "Graph-Based Knowledge Representation and Algorithms for Air and Maintenance Operations." ICAS Congress Proceedings, ICAS, 2022, https://www.icas.org/icas_archive/ICAS2022/data/papers/ICAS2022_0265_paper.pdf .

Lineage appears automatically as edges. Quality assertions trigger alerts and remediation workflows. Access rules evaluate at query time based on user attributes and declared purpose. The result is faster delivery with fewer surprises, since the rules that matter live where the data moves.

Measuring Architectural Health

What gets measured improves. Treat the nervous system like a product with its own indicators:

- **Interoperability score:** percentage of critical concepts mapped to shared semantics.
- **Time to integrate:** median time to onboard a new source or consumer.
- **Policy coverage:** proportion of sensitive datasets with machine-readable rules.
- **Lineage completeness:** fraction of high-value flows with end-to-end traceability.
- **Reuse ratio:** number of consumers per published data product or API.
- **Retrieval effectiveness:** precision and recall of semantic search over the catalog and graph.

These metrics indicate whether the architecture creates flow or friction.

Anti-Patterns to Avoid

- Treating the warehouse or lake as the architecture rather than one component of it.
- Hand-coded point-to-point integrations that bypass shared semantics and metadata.
- Policies written only as documents that humans must interpret on every hop.
- A single global model that erases domain context in the name of uniformity.

- Technology choices that hide relationships that the mission needs to see.

Getting Started without Boiling the Ocean

Begin with your most valuable questions, then model the concepts and relationships that answer them. Build a thin semantic layer around a small set of critical entities. Connect two or three high-value systems through the graph and attach one or two relevant policies. Publish these as reusable data products with clear contracts. Expand outward in rings, guided by real use cases and measured by the indicators above. The goal is steady, compounding interoperability, not a grand rewrite.

A modern enterprise learns and acts through its data architecture. Treat it as the nervous system it is. Invest in semantics, lean into graph thinking, and let interoperability become the habit that makes every other part of modernization easier, faster, and safer.

TREND SNAPSHOT
Graph Thinking Goes Mainstream

- **Knowledge Graph Adoption:** A Gartner survey in 2023 estimated that over 50 percent of enterprises pursuing AI initiatives had already implemented or were piloting knowledge graphs as a foundational enabler. (Gartner 2023[50])
- **Human and Machine Alignment:** Cognitive scientists note that human memory works through associative links, while AI models rely on vector similarity and embeddings. Graphs provide the missing bridge by making relationships explicit and machine-actionable. (Collins 1975[51], Bommasani 2021[52])
- **Interoperability Pressure:** In large enterprises, fewer than 20 percent of critical workflows rely on a single system of record.

50. Gartner. Market Guide for Active Metadata Management. Gartner, Inc., Sept. 2023.
51. Collins, Allan M., and Elizabeth F. Loftus. "A Spreading-Activation Theory of Semantic Processing." Psychological Review, vol. 82, no. 6, 1975, pp. 407–428.
52. Bommasani, Rishi, et al. On the Opportunities and Risks of Foundation Models. Stanford Institute for Human-Centered AI, Aug. 2021.

Most depend on stitching together data from multiple platforms. Graph-based architectures and semantic layers have become the dominant method for enabling that stitching. (IDC 2023[53])

- **Government Interest Rising**: Programs across defense and intelligence are exploring knowledge graph-centric architectures to handle complex questions such as supply chain risk, entity resolution, and cross-domain mission planning. These efforts highlight that interoperability is not an abstract benefit but a mission-critical requirement. (ODNI 2019[54], DoD 2020[55])

KEY TAKEAWAY: Graph thinking is no longer niche. It has become the primary design pattern for organizations that need to connect fragmented data and accelerate AI adoption.

Edge and Cloud Integration

One of the most enduring challenges in technology modernization is deciding how to distribute workloads across environments. The choice is rarely binary. Modern architectures depend on a continuum that blends the global reach and elasticity of the cloud with the immediacy and autonomy of the edge. The two are not competitors but collaborators; each addresses a different part of the performance, security, and mission equation.

The Case for a Hybrid Foundation

Cloud environments have redefined what scale and accessibility mean. Teams can provision resources in minutes, leverage prebuilt AI and analytics services, and collaborate globally through shared platforms. This elasticity allows innovation to move faster than ever before. Yet, in parallel, edge computing has become indispensable in contexts where time, bandwidth, or connectivity cannot be taken for granted. In a tactical operation, an aircraft or autonomous system cannot wait for a round-trip to a distant

53. IDC. Worldwide Knowledge Graphs Forecast, 2023–2027. International Data Corporation, Dec. 2023.
54. Office of the Director of National Intelligence. The AIM Initiative: Augmenting Intelligence Using Machines. ODNI, 2019.
55. U.S. Department of Defense. Data Strategy. Chief Data Officer, Department of Defense, Sept. 2020.

data center. The edge brings computation to the source of action, ensuring continuity even when disconnected.

True modernization lies in harmonizing the two. Hybrid architectures allow organizations to train AI models in the cloud, deploy them at the edge for inference, and then feed results back for continuous improvement. This dynamic exchange (data flowing out, insights flowing back) creates a resilient system that can scale globally while responding locally.

For defense and intelligence organizations, this balance is particularly vital. The enterprise cloud provides analytic depth and shared context; the edge provides speed and survivability in contested environments. The future of modernization will belong to those who can move seamlessly between the two, orchestrating them as one integrated ecosystem. The key is finding the right blend of components and workflows for the mission at hand.

In practice, this means defining which tasks can tolerate latency and which cannot. Bulk data processing, long-term storage, and enterprise-wide analytics are often best suited for the cloud. Real-time decision-making, tactical analytics, and security enforcement at the point of action frequently require the edge. The most successful architectures are designed for seamless handoff between the two, allowing workloads to move dynamically as conditions shift.

Evaluation Principles for Balance

Organizations looking to establish the right balance between cloud and edge should consider several guiding principles:

1. **Mission Criticality**: Identify which functions are too important to risk latency or disruption. These belong at the edge.
2. **Data Sensitivity**: Consider what data can leave the operational environment and what must remain local due to classification, privacy, or regulatory restrictions.
3. **Bandwidth and Connectivity**: Assess network availability and reliability. If connections are intermittent or contested, the edge must carry more responsibility.

4. **Elasticity Requirements**: Evaluate where scale-out capacity is needed. Large-scale analytics and training of AI models are often cloud-centric, while inference at the point of action may run at the edge.
5. **Cost Efficiency**: Balance the economics of running workloads locally versus centrally. Sometimes keeping compute at the edge saves bandwidth costs, while other times consolidating to the cloud reduces duplication.
6. **Policy and Governance**: Ensure architectures can enforce rules consistently across both environments. This requires unified identity, access, and data policies that travel with the workload.

Industry and Government Lessons

Commercial firms increasingly pursue "cloud-to-edge continuums," where the same development and deployment pipelines target both environments. Automotive manufacturers, for example, process sensor data on vehicles in real time but synchronize summaries to the cloud for fleet-wide analytics. Energy companies run AI models at offshore rigs for safety monitoring while uploading periodic insights to central systems for regulatory reporting.

In defense, projects such as the Joint Warfighting Cloud Capability (JWCC) and the adoption of tactical edge devices reflect the same hybrid philosophy. The DoD recognizes that centralized clouds are essential for enterprise analytics and AI training or preparation, yet edge autonomy is indispensable for forward operations. The Intelligence Community faces similar pressures, balancing the need for centralized analytic horsepower with the reality that operators often work in disconnected or compartmentalized environments.

Designing for Fluidity

The ultimate goal is not to decide once and for all where workloads belong. The goal is to design for fluidity. Architectures should enable workloads to shift between cloud and edge as conditions dictate. A sensor may process data locally, then push summaries to the cloud when bandwidth becomes available. A mission planner may run models in a central region during peacetime, then deploy lightweight versions to the edge during contested operations.

Hybrid architectures also require cultural adaptation. Teams must learn to design, deploy, and secure applications across multiple environments. Governance must evolve to ensure policies are enforced consistently. Procurement strategies must allow for flexibility, not lock-in to a single vendor or paradigm.

QUICK NUMBERS
Cloud-to-Edge Adoption Across Sectors

- **72%** of enterprises report using a hybrid cloud model that incorporates some form of edge computing (Flexera 2024 State of the Cloud Report[56]).
- **55%** of industrial firms (manufacturing, energy, logistics) have deployed edge solutions for real-time monitoring and predictive maintenance (IDC 2023[57]).
- **68%** of telecom providers are investing in edge infrastructure to support 5G and low-latency applications (GSMA 2023[58]).
- **43%** of healthcare organizations report using edge for clinical data capture and medical imaging analysis, with another **30%** planning deployments in the next two years (Frost & Sullivan 2023[59]).
- **DoD projection**: By 2028, more than **60%** of AI inference workloads for defense missions are expected to run at the tactical edge, with training and aggregation continuing in the cloud (DoD AI Strategy updates, 2023[60]).

56. Flexera. 2024 State of the Cloud Report. Flexera Software LLC, 2024, https://info.flexera.com/CM-REPORT-State-of-the-Cloud .
57. International Data Corporation (IDC). Worldwide Edge Spending Guide: Forecast for 2023–2027. IDC, Dec. 2023.
58. GSMA. The Mobile Economy 2023. GSM Association, 2023, https://www.gsma.com/mobileeconomy/ .
59. Frost & Sullivan. Global Healthcare Edge Computing Growth Opportunities. Frost & Sullivan, 2023.
60. U.S. Department of Defense. Responsible Artificial Intelligence Strategy and Implementation Pathway. Chief Digital and Artificial Intelligence Office, June 2023.

KEY TAKEAWAY: Hybrid is the norm, not the exception. Organizations across sectors are leaning into the cloud for scale and advanced services, while investing heavily at the edge for immediacy, autonomy, and resilience.

Lessons from the Field

The organizations that succeed at modernization apply the four principles introduced earlier as daily habits rather than abstract models. Their architectures evolve continuously, their procurement cycles encourage experimentation, and their leaders focus as much on human adoption as on technical integration. The examples that follow illustrate how these ideas translate into field-tested results.

Planned Obsolescence as Strategy

One of the most important lessons is that every system has a shelf life. Too many organizations cling to legacy systems long after they should have been retired, pouring resources into sustaining platforms that drain innovation budgets. By contrast, forward-leaning organizations design with obsolescence in mind. They accept from the start that a platform, a service, or even a data model will eventually need to be replaced.

Planned obsolescence is not wasteful; it is strategic. It means building architectures that can absorb change without disruption, documenting exit strategies for critical systems, and budgeting for retirement as deliberately as for adoption. In practice, this can mean adopting containerized workloads that can be ported between providers, structuring contracts with clear off-ramps, or instituting refresh cycles that align with mission milestones rather than arbitrary fiscal calendars.

Modular Architecture as an Operational Principle

Both industry and government have learned that large, centralized systems rarely age well. They tend to accumulate complexity, lock in obsolete practices, and stifle innovation. The lesson is that modularity is more than a design preference; it is an operational principle.

Successful organizations break systems into small, composable services that can be upgraded or replaced independently. They invest in shared interfaces and semantic layers so that new components can be plugged in without rewriting the entire system. This approach mirrors successful strategies in manufacturing, where modular design enables rapid substitution and upgrade without halting the production line. In the technology domain, modularity ensures that modernization is continuous rather than episodic.

Rapid Procurement and Experimentation at Scale

Industry leaders have long relied on agile procurement practices to stay ahead of competitors, and government innovators have begun to catch up. Programs such as the Air Force's Kessel Run and the Navy's Black Pearl have demonstrated that iterative development and rapid acquisition can deliver mission-ready solutions far faster than traditional cycles.

The operational lesson is that procurement must not be viewed as a gate to modernization, but as an enabler of continuous experimentation. The best organizations establish pathways to quickly bring in new tools, test them in realistic conditions, and scale only those that prove their value. They treat procurement not as a single transaction but as a recurring rhythm of piloting, evaluating, and evolving.

Responsive Design for Changing Missions

Modernization is not only about technology; it is about anticipating change. Markets shift, threats evolve, and missions expand. Infrastructure that cannot adapt to these realities quickly becomes a liability. Responsive design focuses on flexibility, scalability, and resilience in the face of uncertainty.

In industry, this might mean designing customer platforms that can rapidly integrate new payment systems, support global expansion, or comply with emerging regulations. In government, it means building architectures that can shift from peacetime analysis to wartime command, or from domestic missions to coalition operations, without requiring wholesale replacement. Responsive design accepts that requirements will change and builds systems that can accommodate that change with minimal disruption.

Federation over Centralization

A recurring misstep in both sectors has been the pursuit of all-in-one solutions that promise to unify everything under a single platform. History shows that such monoliths rarely deliver. They collapse under their own complexity, are slow to adapt, and often alienate users who feel forced into workflows that do not match their needs.

The lesson is that federation often beats centralization. By enabling multiple systems to interoperate through shared standards and semantic models, organizations achieve the benefits of integration without the fragility of uniformity. Federation acknowledges that different domains have unique needs, yet ensures that their outputs can be shared and combined when required.

Continuous Governance and Embedded Security

Another field lesson is that governance and security cannot be afterthoughts layered onto architectures after they are built. They must be embedded into the design from the beginning. Continuous governance means that policies are not written documents on a shelf but are operationalized in systems as machine-readable rules. Embedded security means that authentication, authorization, and monitoring are woven into the architecture's nervous system, not bolted on as external checkpoints.

The organizations that do this well treat governance and security as accelerants rather than inhibitors. By embedding rules into the architecture, they allow modernization to proceed faster because compliance is automatic and risk is reduced.

Measuring Modernization Progress

What cannot be measured cannot be managed. Organizations that succeed at modernization treat measurement as part of the architecture itself. They track indicators such as:

- Ratio of investment in new capabilities versus legacy sustainment
- Time to integrate new data sources or applications
- Percentage of systems designed with modular or containerized components

- Adoption and satisfaction rates among end users
- Number of security incidents prevented through embedded controls

By monitoring these indicators, leaders can tell whether modernization is truly delivering value or simply reshuffling budgets.

Building for Decades, Not Just Today

The final and perhaps most important lesson is that modernization must be approached as a generational investment. Decisions made today will shape an organization's trajectory for decades. That does not mean committing to a single platform or vendor, but rather committing to principles that will endure: modularity, interoperability, federation, and continuous governance.

Industry giants that have lasted for decades, from Amazon to Microsoft, have done so by continuously reinventing their infrastructure. Government programs that succeed, such as the National Geospatial-Intelligence Agency's modernization efforts, are those that plan not just for the current mission but for the missions that will emerge in five, ten, or twenty years.

An Operational Blueprint

Taken together, the lessons from the field point to a clear operational blueprint:

1. Accept obsolescence as inevitable and design for graceful retirement.
2. Favor modular, composable architectures over monoliths.
3. Treat procurement as a continuous enabler of experimentation, not a hurdle.
4. Build for a responsive design that anticipates change.
5. Embrace federation to balance autonomy with interoperability.
6. Embed governance and security as part of the architecture.
7. Measure progress relentlessly and adjust course based on results.

The organizations that follow this blueprint are not merely modernizing their technology; they are modernizing their ability to modernize.

They are creating infrastructure that can evolve with the market, the mission, and the unknowns of the future. That is the essence of technology modernization done right.

VOICES IN THE FIELD
Modernization in Practice

- **Thomas Davenport (author, *Competing on Analytics*)**: Davenport emphasizes that modernization is never static. He notes that organizations that succeed are those that "treat technology as a renewable resource, constantly refreshed and realigned to business outcomes rather than allowed to ossify."
- **General David L. Goldfein (former USAF Chief of Staff)**: In discussing digital modernization, Goldfein highlighted the need for continuous adaptation, arguing that the Air Force must "build systems designed to evolve with the fight, not systems built for the last fight."
- **Satya Nadella (CEO, Microsoft)**: Nadella has repeatedly pointed out that "every company is now a software company," meaning modernization is not the responsibility of IT alone but a core business and mission concern. His guidance: architectures must empower every part of the enterprise, not just central technology functions.
- **U.S. GAO Reports on Federal IT**: Oversight agencies have noted that legacy IT consumes the majority of federal IT budgets, crowding out innovation. The GAO repeatedly stresses that "measuring and managing technical debt is as essential as tracking financial liabilities."
- **Admiral William H. McRaven (ret.)**: Speaking on operational agility, McRaven observed that "speed and adaptability are the new currency of advantage." His point applies directly to modernization: architectures that cannot adapt will not keep pace with mission demand.

KEY TAKEAWAY: Across domains, the consistent refrain is that modernization is not about a single technology choice. It is about designing for evolution, treating adaptability and measurement as non-negotiables, and embedding modernization into the culture of the organization.

Building Living Architecture

Architecture is not a set of diagrams placed in a binder or a static artifact that can be signed off on once and then left alone. It is a living practice that must evolve in step with the organization, the mission, and the technology landscape. The pace of change in both industry and government is simply too fast for rigid structures to survive. New threats emerge, markets shift, and innovations become mainstream in a fraction of the time it once took. In this environment, the organizations that thrive are those that treat architecture as a continuous process rather than a one-time design.

Architecture as Culture, Not Just Design

To succeed, architectural evolution must be embedded in the organization's culture. This means leaders do not merely build systems; they cultivate habits. Teams learn to expect change and plan for it. Processes are designed assuming that components will be replaced, upgraded, or retired. Funding models acknowledge that modernization is ongoing, not episodic. The culture that emerges views architecture less as a structure to defend and more as a landscape to steward.

When cultural alignment is absent, organizations fall into the trap of treating architecture as a monument. They invest heavily in a design, declare victory once it is built, and then resist modification even as requirements shift. The result is brittle infrastructure that ages poorly and eventually demands costly overhauls. By contrast, a culture of living architecture emphasizes resilience through adaptability, valuing the ability to change over the illusion of permanence.

The Process of Continuous Evolution

Living architecture also requires a disciplined process. Continuous review cycles become part of the operational rhythm. Instead of waiting for multi-year modernization programs, organizations establish quarterly or even monthly checkpoints to examine whether systems still serve their purpose. Emerging technologies, shifting missions, and evolving user needs are evaluated against existing designs. Adjustments are made incrementally, ensuring that architecture never falls too far out of sync with reality.

Responsive feedback loops are critical. Users must have avenues to signal when systems no longer meet their needs. Architects and engineers must be empowered to propose modifications, and leadership must be prepared to allocate resources for iterative improvements rather than insisting on perfection before delivery. In this way, architecture begins to mirror agile software development practices, with small adjustments compounding into meaningful transformation over time.

Connecting Tactical Lessons to Living Practice

The field lessons discussed earlier, planned obsolescence, modular design, rapid procurement, and responsive architectures, all converge in the idea of living architecture. Planned obsolescence ensures that systems are not clung to beyond their utility. Modularity ensures that components can be swapped out without destabilizing the whole. Rapid procurement ensures that new capabilities can be brought in as soon as they are needed. Responsive design ensures that architectures remain aligned with changing missions.

Taken together, these practices shift architecture from a static state to a continuous motion. The enterprise's nervous system is always adapting, recalibrating, and positioned for the next challenge.

The Human Dimension

Just as important as process is the human dimension. Living architecture requires that people across the enterprise see themselves as participants in its evolution. Architects, engineers, operators, and end users must all understand that their input matters and that they share responsibility for shaping the systems they rely on. Leaders must reinforce this mindset by telling the story of modernization not as a project imposed from above, but as a shared practice that empowers everyone.

A Bridge to the Future

By treating architecture as living, organizations move beyond the cycle of periodic overhauls that consume resources and disrupt operations. Instead, modernization becomes habitual. The system evolves continuously, aligning with the rhythm of mission needs and technological innovation. This does not eliminate challenges, but it ensures that organizations are never caught completely unprepared.

A living architecture is not the absence of design; it is the elevation of design into a permanent discipline. It is the recognition that success comes not from building a perfect system once, but from building the capacity to evolve systems forever. This mindset creates the bridge between the tactical lessons of today and the aspirational vision of tomorrow, where technology and mission advance together in harmony.

LOOKING AHEAD
The Future of Living Architecture

- **Self-Healing Systems**: Architectures will increasingly incorporate AI-driven monitoring and remediation, allowing systems to detect issues, reconfigure themselves, and recover without human intervention.
- **Policy as Code at Scale**: Governance will evolve from human-readable directives to fully embedded, machine-executable rules that travel with data and workloads across every environment.
- **AI-Orchestrated Workflows**: Intelligent orchestration layers will dynamically move workloads between cloud and edge based on mission context, bandwidth, and cost considerations–blending environments seamlessly.

- **Semantic Interoperability as the Norm**: Knowledge graphs and semantic layers will become standard, ensuring that systems not only exchange data but also share meaning.
- **Composable Mission Systems**: Defense and intelligence organizations, along with industry leaders, will adopt modular systems that can be reassembled quickly for new missions, resembling "mission kits" of capabilities.
- **Continuous Experimentation Culture**: Modernization will no longer be episodic. Organizations will normalize constant piloting, evaluation, and integration of new tools as a way of working, not a special initiative.

KEY TAKEAWAY: The future of living architecture is not about a final state. It is about building the processes, culture, and mindset that allow organizations to evolve endlessly. The most successful enterprises will be those that stop chasing modernization as a project and start living it as a permanent discipline.

Capturing Hearts and Minds

The most overlooked aspect of modernization is the human one. Technology may be procured, deployed, and integrated, but its real test comes in the hands of the people who use it. Without their trust, enthusiasm, and engagement, even the most advanced systems will falter. Modernization, at its core, is not a purely technical endeavor; as mentioned before, it is a socio-technical challenge. Success requires not only building the right architectures but also creating the social conditions where people embrace and champion change.

Modernization as a Social Contract

Every modernization effort asks something of the workforce: to learn new systems, to adopt new workflows, or to rethink familiar habits. In return, the workforce expects something as well: that the technology will make their lives easier, their work more meaningful, and their contributions more visible. When modernization is framed as a social contract, trust is built. People see that the organization is not imposing change for its own sake but investing in tools that serve the mission and the people equally.

Engaging the "People Part" of the System

A socio-technical view of modernization recognizes that people are not passive recipients of technology. They are co-creators of how systems are used, adapted, and improved. Successful organizations engage people early and often, weaving them into the design process rather than presenting them with finished solutions. This can take many forms:

- **User councils** that test prototypes and provide feedback before wide deployment.
- **Shadowing exercises,** where designers and engineers observe workflows directly to understand the real challenges users face.
- **Cross-functional workshops** that bring together operators, technologists, and policy experts to surface concerns and build shared understanding.

These practices shift modernization from being something "done to" the workforce to something "done with" the workforce.

Building Champions and Advocates

One of the most powerful accelerators of modernization is the emergence of internal champions. These are the early adopters who see the value of new tools and help their peers understand and embrace them. Leaders can intentionally cultivate champions by identifying motivated individuals, giving them early access to technology, and empowering them to share their experiences. When peers hear from trusted colleagues rather than distant executives, adoption accelerates.

Recognition plays a role as well. Highlighting success stories, rewarding innovative uses of new systems, and publicly celebrating contributions signal to the organization that embracing modernization is valued and visible. These acts build momentum, turning early adopters into advocates and advocates into cultural change agents.

Storytelling and Mission Alignment

Change is more readily accepted when it is clearly connected to purpose. Modernization must be framed not as an IT upgrade, but as a mission enabler. Storytelling is one of the most effective ways to make this connection. When people hear concrete stories about how a new system reduced errors, sped up decision-making, or enabled collaboration across boundaries, the abstract idea of modernization becomes tangible.

Leaders who tell these stories authentically, tying technology directly to outcomes that matter in daily work, create alignment between personal effort and organizational goals. That alignment turns compliance into commitment.

Overcoming Fear and Resistance

It is natural for people to resist change, especially when it threatens established expertise or creates uncertainty about roles. Successful modernization efforts acknowledge this resistance openly rather than ignoring it. They provide training that builds confidence, create safe spaces for questions, and offer support for those struggling to adapt.

In some cases, resistance is rooted in legitimate concerns about workload, trust, or unintended consequences. Treating these concerns as valuable feedback rather than obstacles can improve both the technology and the process. When people see that their input changes outcomes, resistance often turns into ownership.

Accelerating the Organization

The social dimension of modernization is not simply about easing adoption. It is about creating alignment that accelerates the entire organization. When people understand the purpose of modernization, see their own fingerprints on the systems being deployed, and are celebrated for embracing new ways of working, the organization moves faster. Innovation becomes less about mandates from above and more about energy from within.

Practical Examples of Engagement

Several practices consistently prove effective in capturing hearts and minds:

- **Pilot with the willing**: Start with teams that are motivated to try new solutions and showcase their successes as models for others.
- **Design for quick wins**: Deliver small improvements that demonstrate immediate value, building credibility for larger changes.
- **Communicate relentlessly**: Use multiple channels to explain not only the "what" but also the "why" of modernization.
- **Invest in learning**: Provide structured training, on-demand resources, and peer-to-peer mentoring to ease transitions.
- **Celebrate progress**: Share stories of how new tools improved outcomes and highlight individuals who made the transition possible.

From Compliance to Commitment

Ultimately, capturing hearts and minds is about shifting from compliance to commitment. Compliance ensures that people use the tools because they must. Commitment ensures that they use the tools because they believe in them. Compliance produces reluctant adoption; commitment produces champions who drive adoption forward.

When modernization is approached as a socio-technical challenge, it ceases to be a battle between technology and people. Instead, it becomes a shared project of building systems and cultures that evolve together. That alignment, between architecture and workforce, between process and purpose, is what accelerates organizations into a future where modernization is not resisted, but embraced.

Outdated Models to Leave Behind

As with any discipline, technology modernization has its share of experiments, trends, and approaches that once held promise but have since proven insufficient. They were born from real needs and often represented the best thinking of their time, yet over the years, their limitations became clear. Understanding these models in depth (their origins, their temporary successes, and their eventual obsolescence) provides a critical perspective. It prevents organizations from repeating old mistakes while preserving the valuable lessons embedded within them.

The Era of Big-Bang Deployments

For decades, organizations attempted to modernize through large-scale, "big-bang" deployments. These efforts were grounded in the belief that a sweeping transformation could be achieved by designing a complete solution, spending years developing it, and then unveiling it all at once. The intent was noble: to leap forward in one decisive move rather than inching forward with incremental changes. In certain contexts, particularly when technology cycles were slower and requirements more stable, this approach occasionally succeeded.

Yet as environments grew more dynamic, the flaws became glaring. By the time a large program was ready for deployment, the mission had often shifted, the technology had become outdated, or user expectations had outpaced what was delivered. Budgets ballooned, delays mounted, and what was intended as a transformation became, instead, a monument to inertia. What remains valuable from this model is its ambition; the recognition that modernization requires vision and boldness. But ambition must now be paired with agility. Incremental, iterative change is far more aligned with the pace of both technology and mission evolution.

Centralized "One-Stop Shop" Systems

Another outdated model is the pursuit of massive, centralized platforms designed to be the single system for every enterprise function. These "one-stop shops" promised efficiency through consolidation. Leaders were enticed by the prospect of uniformity: one system to train on, one interface to support, one vendor relationship to manage. For a time, such systems seemed appealing, especially in organizations struggling with fragmentation and silos.

However, these monoliths often collapsed under their own weight. They were rigid, difficult to customize, and rarely kept pace with diverse and evolving user needs. In practice, instead of eliminating silos, they often created new ones within the giant system itself. Worse still, users felt forced to conform their workflows to the system, rather than the system adapting to support the mission. The valuable lesson to preserve here is the desire for interoperability and integration. The problem was not the goal but the method. Today, federation and modularity achieve the same end: unified outcomes without the fragility of uniform systems.

Technology-First Modernization

A third model that has repeatedly failed is modernization driven by the allure of technology itself. Organizations sometimes pursued modernization by chasing the newest tools or platforms simply because they were new. The thinking was that by adopting cutting-edge technology, the organization would automatically become modern. Vendors often reinforced this mindset, offering polished promises that their solution alone could transform operations.

While this approach sometimes injected useful tools into the enterprise, it often led to adoption without alignment. Systems were introduced that did not connect to the mission, did not integrate with existing infrastructure, and did not resonate with users. The result was a graveyard of unused applications and wasted investment. The insight worth salvaging here is the spirit of curiosity and willingness to explore new tools. Innovation requires experimentation. Yet the lesson is clear: technology must serve outcomes, not the other way around. Mission-centered design is the only reliable anchor.

Paper-Centric Governance

For years, governance was treated as a compliance exercise bound in paper. Policies were drafted, reviewed, and filed in binders or digital archives, rarely operationalized into actual practice. The assumption was that if the rules were written down, they would be followed. In reality, these static artifacts created the illusion of control while leaving systems vulnerable to drift and inconsistency. Users often ignored the rules out of necessity, improvising solutions when the formal guidance proved impractical.

This model has been left behind because it fails to keep pace with modern systems and threats. Governance today must be automated, embedded, and actionable. What is worth preserving is the intent behind the old model: clarity, accountability, and formal authority. The future of governance retains those qualities while ensuring that rules live inside the architecture itself, as machine-readable policies that travel with data and enforce themselves in real time.

The Outsourcing Imperative

Another once-popular approach was to outsource modernization wholesale to external vendors. Leaders assumed that modernization was primarily a technical problem and that expertise resided outside the organization. Vendors were brought in to design and deliver solutions, often with little engagement from the workforce who would ultimately use them. While outsourcing brought short-term capacity, it often failed to build internal capability. Organizations became dependent on external actors, unable to sustain or evolve their systems once contracts ended.

This model has been replaced by more collaborative partnerships that balance external expertise with internal ownership. Modern organizations may still work with vendors, but they insist on co-creation, shared responsibility, and skill transfer. The enduring lesson is that outside expertise is valuable, but modernization cannot be entirely outsourced. It must be owned by the people whose mission it serves.

Why These Models Failed

Each of these models emerged from real needs: the desire for bold transformation, the search for integration, the pursuit of innovation, the demand for governance, and the reliance on expertise. None were misguided in intent. What failed was the assumption that modernization could be static, centralized, or detached from mission and people. The common thread across their shortcomings is rigidity: the belief that a single design, tool, or vendor could deliver modernization once and for all.

Toward What Endures

The organizations that are succeeding today are those that take the best from these outdated models and discard the rest. They retain the ambition of big-bang transformations but achieve it through iterative delivery. They preserve the desire for integration without falling into the trap of monoliths. They channel curiosity about new technology into focused, mission-driven experimentation. They honor the need for governance but embed it directly into systems. They value external expertise while insisting on internal ownership.

The failures of the past are not wasted. They are the scaffolding for more durable approaches. By studying them honestly, leaders are reminded that modernization is not a one-time effort or a silver bullet. It is a continuous practice of aligning people, process, and technology in ways that are flexible, sustainable, and mission-relevant. Leaving outdated models behind clears the path toward architectures and cultures that are not only modern today but prepared to evolve for decades to come.

INSIGHT
From Lessons of the Past to the Rhythm of the Future

Understanding the shortcomings of past approaches is not about assigning blame, but about building wisdom. Each failed model reveals a trap to avoid, and each surviving element offers a foundation to build upon. What emerges is a clear truth: modernization is not a project with an end date, but a rhythm to be cultivated.

The future belongs to organizations that see modernization as an everyday practice, woven into the culture as naturally as budgeting, training, or operations. When technology, people, and process are allowed to evolve together in continuous motion, modernization ceases to be an episodic disruption and becomes an enduring capability. The next step, therefore, is to reimagine modernization not as a destination, but as a way of working; an everyday discipline that ensures relevance, resilience, and readiness.

Modernization as a Way of Life

Modernization cannot remain something extraordinary, something that happens in large bursts followed by long periods of stasis. The lesson from decades of attempts, successes, and failures is that true modernization only takes root when it becomes part of the organizational rhythm. It must be as routine as financial planning, as ingrained as training cycles, and as expected as annual readiness assessments. In other words, modernization must become a way of life.

When architecture and technology practices are embedded into the culture, they no longer feel like disruptive projects forced upon the workforce. Instead, they become the background hum of progress. Systems evolve continuously, tools improve incrementally, and new capabilities arrive in measured steps that feel natural rather than jarring. This shift turns modernization from episodic to habitual, a permanent discipline that sustains relevance and agility in a world that never stops changing.

From Episodic Change to Continuous Renewal

The organizations that succeed are those that abandon the illusion of permanence. They know that every system has an expiration date, every integration will need to be reworked, and every policy will need to be reinterpreted. Instead of resisting these cycles, they embrace them. They design for change. They fund refresh as deliberately as they fund sustainment. They treat each iteration not as an emergency fix but as part of a planned rhythm of renewal.

This cultural shift unlocks energy. People no longer brace for modernization as disruption. They come to expect it as an opportunity. Engineers anticipate cycles of improvement. Users trust that their needs will be heard in regular feedback loops. Leaders stop talking about "the next big transformation" and start talking about "how we continuously evolve."

The Role of the CDAO in Everyday Modernization

For the Chief Data and AI Officer, the challenge is not to own modernization as a single project, but to steward modernization as an enduring practice. This means establishing the frameworks and habits that keep the architecture alive. It means weaving together processes that ensure governance, security, and interoperability are never static but always responsive. Most importantly, it means telling a story of modernization that resonates with people across the organization; not just technologists, but operators, analysts, leaders, and partners.

By embodying this mindset, the CDAO helps move the organization from compliance to commitment. People no longer adopt new systems reluctantly; they engage with them eagerly because modernization is seen as a shared journey that advances both mission and individual purpose.

Modernization as a Cultural Strength

Ultimately, modernization as a way of life is not about technology at all. It is about culture. It is about an organization that values adaptability, rewards curiosity, and builds resilience by constantly renewing itself. This culture does not fear obsolescence because it has already accepted that systems will change. It does not fear complexity because it knows interoperability and federation are part of the design. It does not fear disruption because modernization has become a normal, expected part of how work gets done.

Such a culture is prepared not only for today's challenges but for the unknowns of tomorrow. It will not be caught flat-footed by shifts in technology, mission, or market because it is already in motion. The heartbeat of everyday modernization will carry it forward.

A Closing Invitation

This chapter has examined frameworks, lessons, outdated models, and the living practice of architecture. The common thread is clear: modernization cannot be an occasional initiative. It must be a permanent capability. The organizations that embrace this truth will not only stay ahead of change but will help define it. They will set the pace rather than struggle to keep up.

The invitation to every leader, technologist, and user is to see modernization not as something imposed, but as something shared. To see it not as a disruption, but as a renewal. And to live it not as an obligation, but as an advantage. When modernization becomes a way of life, organizations do not just survive the future; they shape it.

Chapter 7

AI & ANALYTICS STRATEGY

Artificial intelligence and advanced analytics have become central to how modern organizations generate insight, create value, and maintain relevance in a rapidly shifting environment. For leaders at every level, from Chief Data Officers and Chief Analytics Officers to Vice Presidents of technology and mission operations, the mandate is clear: these capabilities are no longer optional enhancements. They are core to organizational competitiveness and mission success. Yet despite this urgency, the concept of an AI strategy remains frustratingly nebulous for many. Too often, it becomes a placeholder term, shorthand for purchasing expensive technology platforms, chasing vendor-driven hype cycles, or launching high-profile pilots that never progress beyond glossy slides and conference presentations.

The reality is that AI and analytics demand far more than a budget line item or a shiny new tool. They require a disciplined, flexible approach that treats these capabilities as socio-technical systems—a theme that runs throughout this book. Algorithms, infrastructure, governance, workflows, and people are interconnected components that rise or fall together. Success is not only about model accuracy or the speed of data pipelines. It is about embedding these tools into the organization's culture and processes so that insights flow naturally into decisions, risks are surfaced early, and human judgment is consistently augmented rather than replaced.

The challenge for today's leaders is twofold. On one hand, they must inspire a vision of what AI and analytics could make possible for their enterprise, whether that is accelerating research, reducing operational costs, strengthening national security, or improving citizen services. On the other hand, they must confront the complex operational realities of capability evaluation, workforce readiness, governance requirements, and cultural adoption. A vision that ignores execution is an empty aspiration, while an execution plan without vision risks becoming a series of disconnected projects with no lasting impact. A modern AI and analytics strategy requires the synthesis of both.

This chapter is designed to meet leaders at that intersection. It offers a structured approach to making AI and analytics real: moving from pilot to portfolio, from experimentation to sustainment, from novelty to necessity. Unlike the broader modernization themes explored earlier, which established the technical foundations for future capabilities, this chapter turns to the operational craft of strategy itself. It focuses on the challenging yet essential task of integrating AI and analytics into organizations' everyday operations. For a Chief Data and AI Officer, this is the role's center of gravity. For executives across the enterprise, this is where lofty ambitions about data-driven transformation either translate into measurable impact or collapse under their own weight.

Knowing Where You Stand

The path forward begins with understanding your current position. No organization can chart a credible course for AI and analytics without first taking a clear-eyed inventory of its own strengths, weaknesses, and blind spots. Before pilots are launched, before portfolios are built, before scaling becomes possible, leaders must confront the current reality of their data readiness, technical foundations, organizational processes, and cultural environment. This diagnostic work is not glamorous, and it rarely makes headlines, but it is the bedrock of every durable strategy.

Strategies fail most often not from poor vision, but from misjudged reality. Every organization already possesses a dense ecosystem of systems, habits, and expectations that must be faced before new ambitions take root. Evaluation must therefore begin with a holistic view that encompasses people, technology, process, and culture in equal measure.

People include not only data scientists and engineers but also the analysts, operators, and leaders who must interpret outputs and act on them. Their literacy and willingness to adapt often matter more than the number of experts employed. A team of brilliant data scientists cannot succeed if the decision-makers they serve distrust their work or lack the vocabulary to engage with it. Technology covers both the visible platforms and the invisible plumbing: metadata, APIs, integration points, and the countless small connections that determine whether data can flow where it is needed. The real test is not how modern the stack looks on a vendor slide, but whether it supports transparency, scalability, and auditability in practice. Processes determine reliability. Standardized data intake, validation protocols, and feedback loops create consistency that tools alone cannot provide. And culture, though hardest to measure, may be most decisive of all. A culture that hides failure or works in silos will stall even with advanced technology, while one that rewards learning and collaboration can progress with less.

Internal assessment must be paired with an honest reading of the market context. Technologies commoditize quickly; what was frontier yesterday may be table stakes today. Cloud-based machine learning services have dramatically lowered the barrier to entry into predictive analytics. Federated learning approaches now make once-intractable data sharing problems manageable. Mapping internal reality against these external benchmarks prevents both complacency and overreach. It reveals where you are behind and must catch up, as well as where you have built advantages worth protecting.

Every evaluation must also define desired outcomes with precision. The goal is not AI for its own sake, but to improve decision-making and mission delivery. Whether faster threat analysis, reduced administrative burden, or better customer experience, clarity about purpose keeps ambition grounded. Overreach occurs when organizations chase technologies that do not align with their mission needs, creating more complexity than value. The most effective CDAOs use this clarity as a filter: every initiative must connect to an outcome that leadership already cares about.

This evaluation is not a one-time event. It is a discipline. Just as financial markets require regular audits and performance reviews, AI and analytics strategies require ongoing reevaluation as technologies evolve, missions shift, and new opportunities emerge. This baseline establishes the foundation for all subsequent decisions in the strategy. Without it, pilots may chase the wrong outcomes, portfolios may be imbalanced, and platforms may be built on fragile foundations. With it, a Chief Data and AI Officer can move from vague aspiration to focused execution, ensuring that the enterprise's ambitions are grounded in what is truly possible.

Confronting the Skeptics

Every AI strategy will face resistance. Some opposition is procedural: budget controllers questioning ROI, procurement teams wary of vendor lock-in, and legal departments concerned about liability. But the most persistent resistance is philosophical. Skeptics question not just the execution but the premise itself. They have seen technology fads come and go. They remember when big data was going to solve everything, when blockchain promised to revolutionize governance, when robotic process automation was marketed as the cure for inefficiency. They have watched as expensive pilots were launched with fanfare, only to disappear quietly when results failed to materialize. Their skepticism is not ignorance. It is pattern recognition.

The CDAO who dismisses these voices as obstructionist or technophobic makes a critical error. Skeptics often raise the questions everyone else is thinking but afraid to ask. Their concerns, properly addressed, become the foundation of a more resilient strategy. The work is not to convert them into believers, but to earn their respect through evidence and transparency.

The most common skeptical positions fall into recognizable categories, each requiring its own response strategy:

"We've tried this before, and it didn't work." This objection reflects institutional memory of failed projects, often initiatives launched without clear objectives or adequate support. The response is not to argue that this time will be different, but to show how it will be different. What specific lessons from past failures have been incorporated? What governance structures now exist that were absent before? What changed in leadership commitment, technical capability, or organizational readiness? Concrete answers build credibility where promises do not. In some cases, this provides the opportunity to give these objectors a voice in the process to help identify, capture, and incorporate learnings from past failures.

"AI is a black box we can't trust." This concern about opacity is legitimate, especially in high-stakes environments where decisions must be explainable. The response is not to defend inscrutability but to demonstrate commitment to transparency. Show how model cards document decision logic, how audit trails capture data lineage, and how human oversight functions for critical decisions. Invite skeptics into pilot reviews where they can see explainability in practice. In the best cases, the tools, systems, and models you create incorporate transparency and auditability measures. Trust grows not from technical lectures about algorithms, but from repeated demonstrations that the organization takes transparency seriously.

"This will eliminate jobs." Fear of displacement is perhaps the most emotionally charged objection, and it cannot be dismissed with platitudes about augmentation versus replacement. The honest response acknowledges that some roles will change fundamentally, while being clear about the organization's commitment to reskilling and transition support. Share concrete examples of how AI has expanded rather than contracted opportunity in pilot domains: analysts freed from data cleaning to focus on interpretation, customer service representatives empowered with better information, and field technicians equipped with predictive insights. When people see their peers succeeding with AI rather than being replaced by it, anxiety diminishes.

"We don't have the data quality to make this work." This objection often comes from data professionals who understand the gap between the current state and what AI requires. The response is not to promise perfect data, but to show how strategy accounts for imperfection. Explain which use cases are robust to noisy data and which require cleanup first. Demonstrate how pilots include data quality monitoring and how feedback loops drive incremental improvement. Acknowledge that some applications must wait until foundational work is complete, and show that this sequencing is deliberate rather than something you are ignoring.

"The vendors are just selling us expensive products we don't need." This reflects justified wariness of technology hype and aggressive marketing. The response is to separate vendor relationships from capability decisions. Show how requirements were defined before vendor selection, how proofs of concept test claims before commitment, how contracts include performance guarantees and exit clauses. Demonstrate that the organization is driving the strategy, not being driven by it.

"AI introduces risks we can't control." This concern about bias, drift, adversarial attacks, and unintended consequences reflects a genuine understanding of AI's limitations. The response is not to minimize these risks but to show how they are being managed. Walk through the risk register, explain monitoring protocols, describe rollback procedures, and share how the organization has learned from early missteps. Skeptics respect leaders who treat risk seriously more than those who promise risk-free solutions.

The most effective strategy for managing skepticism is to enlist skeptics as advisors rather than viewing them as obstacles. Invite the most articulate critics to serve on governance boards or pilot review panels. Their questions will sharpen your thinking, their presence will signal openness to dissent, and their eventual endorsement (if earned) will carry more weight than that of any natural ally. When a known skeptic says "this is different," the organization listens.

Regular communication reinforces this approach. Share both successes and failures openly. When pilots underperform, explain why and what was learned. When technical debt accumulates, acknowledge it and outline the plan to address it. When adoption lags, investigate the barriers rather than blaming users. This transparency converts skeptics from adversaries into partners in problem-solving.

Finally, remember that some skepticism will never be converted, and that is acceptable. The goal is not unanimous enthusiasm but informed consent and respectful engagement. An organization where everyone agrees without question is one where critical thinking has been suppressed. Healthy skepticism, properly channeled, strengthens strategies.

Strategic Sequencing: The When and How of AI Rollout

AI strategies often fail not because they pursue the wrong capabilities, but because they pursue them in the wrong order. An ambitious leader might push for advanced natural language processing when the organization has not yet mastered basic analytics. A team might attempt to scale a pilot before governance frameworks are in place to support it. Well-intentioned initiatives collide, creating resource conflicts and political tensions that undermine the broader program. Strategic sequencing prevents this chaos by establishing a deliberate progression of capabilities that builds on prior success while managing organizational capacity for change.

Sequencing is not about moving slowly. It is about moving purposefully, ensuring that each stage creates the conditions for the next. Think of it as constructing a building: you can rush and pour the concrete for the tenth floor while the foundation is still curing, or you can sequence the work so that each level stabilizes before the next begins. The latter approach may seem slower initially, but it reaches completion faster because it avoids catastrophic rework.

The foundation layer focuses on data readiness and organizational literacy. Before any AI pilot launches, basic capabilities must be in place: reliable data pipelines, consistent metadata, and a workforce that understands what AI can and cannot do. This stage is often the least exciting, but skipping it guarantees later failure. Organizations that rush into sophisticated models while their data remains scattered, poorly documented, or of questionable quality will spend months troubleshooting problems that could have been prevented. Invest this time in workshops that build data literacy across departments, in cleaning and cataloging priority datasets, and in establishing data governance practices that will support future work. The metric for readiness is not perfection but sufficiency: do we have enough quality and understanding to support a bounded experiment?

The experimentation layer introduces structured pilots that test both technology and adoption patterns. These initiatives should be chosen strategically, not opportunistically. The first pilots set the tone for the entire program. They should be significant enough to demonstrate real value but bounded enough to succeed within reasonable timeframes. They should involve multiple stakeholders to build cross-functional understanding and should target problems that leadership already cares about. A successful early pilot does more than validate a technical approach; it creates proof that the organization can absorb change, that AI delivers on its promises, and that the team leading the effort knows what it is doing. This success becomes the foundation for more ambitious work.

Pilot sequencing follows a logic of expanding scope and complexity. Begin with use cases where data quality is good, outcomes are measurable, and stakeholder support is strong. A demand forecasting model in supply chain management, for instance, offers clear metrics and affects a concentrated group of users. Success here builds credibility for more complex efforts, such as fraud detection or predictive maintenance, where data may be messier and stakeholders more distributed. The most challenging applications, those involving unstructured data, real-time decision-making, or ethically sensitive domains, should come later in the sequence once the organization has developed both technical capability and organizational muscle memory for managing AI.

The scaling layer transitions successful pilots into operational systems. This phase requires different skills and structures than those required for experimentation. Scaling is not simply running a pilot on more data; it involves productionizing code, automating monitoring, establishing support workflows, and integrating AI outputs into existing business processes. Many organizations stumble here because they treat scaling as a technical exercise when it is fundamentally an organizational one. Scaling requires change management, training programs, updated policies, and often new roles. The sequencing principle is to scale only when these supporting structures are in place. A pilot that succeeds in a controlled environment but fails when deployed widely does more harm than good; it erodes confidence and creates resistance to future efforts.

As scaled systems mature, the portfolio layer emerges naturally. With multiple AI capabilities in production, attention shifts from individual projects to portfolio management: balancing risk across use cases, optimizing resource allocation, identifying synergies, and maintaining a pipeline of new experiments while sustaining existing systems. This is where strategic choice becomes most critical. Not every pilot should scale. Not every scaled system deserves continued investment. Portfolio management requires discipline to say no, to retire underperforming assets, and to double down on what works.

Throughout this progression, capacity management is essential. Organizations have finite resources for change: budget, talent, and management attention. Overcommitting in too many directions simultaneously is a common failure mode. Strategic sequencing respects these constraints by establishing a work cadence that aligns with organizational capacity. Some enterprises can support three or four active pilots concurrently; others can barely manage one. Some can scale new capabilities quarterly; others need annual cycles. The right pace is the one your organization can sustain while maintaining quality and learning from each stage.

Timing decisions also respond to external factors. Regulatory changes may accelerate or delay certain initiatives. Market shifts may suddenly prioritize capabilities that were previously lower priority. Budget cycles constrain when investments can be made. Strategic sequencing is not a rigid plan but a flexible framework that accommodates these realities while maintaining overall coherence.

Communication about sequencing is as important as the sequencing itself. Stakeholders need to understand not only what is happening now but also what is coming next and why. When a team advocates for a particular use case but is told it must wait, they deserve an explanation rooted in sequencing logic rather than arbitrary gatekeeping. When a pilot succeeds but is not immediately scaled, users need to understand what preparatory work must occur first. Transparency about sequence and timing builds patience and trust.

Finally, sequencing enables learning across generations of work. Early pilots inform how later ones are structured. Scaling challenges in one domain guide how others are approached. This cumulative learning is one of the most valuable outputs of a well-sequenced strategy. Each phase not only delivers capability but also generates organizational knowledge that makes subsequent phases faster and more successful.

The Functional Mechanics of AI Rollout

Understanding what to do is only half the battle; knowing how and when to do it determines success or failure. The functional mechanics of AI rollout translate strategic intent into operational reality. This is where abstraction meets implementation, where plans encounter the messy complexity of real organizations, and where the CDAO's ability to orchestrate across functions becomes most critical.

An effective rollout begins long before any model is deployed. It starts with stakeholder alignment, ensuring that everyone affected by the initiative understands its purpose, their role, and what success looks like. This is not a single meeting but an ongoing dialogue. Identify executive sponsors who will champion the work and provide top cover when obstacles emerge. Engage middle managers who will need to adjust workflows and performance expectations. Connect with frontline users whose daily work will change. Each audience requires different messages and different levels of detail, but all need to hear a consistent narrative about why the change matters and how it aligns with their interests.

Technical preparation runs in parallel. Infrastructure must be in place to support the new capability: computing resources, data access, integration points, and monitoring tools. Security and compliance reviews should happen before deployment, not after. The worst rollouts are those where technical teams rush to deploy only to discover that privacy reviews or security protocols block progress. Front-loading these conversations prevents later surprises and demonstrates organizational maturity.

Training and enablement determine whether the new capability will actually be used. Investing in sophisticated AI and then expecting users to figure it out on their own is a recipe for failure. Develop tiered training: brief orientations for peripheral stakeholders, detailed workshops for power users, and technical deep-dives for support staff. Create reference materials that people will actually use: quick-start guides, FAQs, scenario walkthroughs, and troubleshooting flowcharts. Make experts available during initial rollout when questions will be most frequent and urgent.

Phased deployment reduces risk and accelerates learning. A full organizational rollout on day one invites disaster. Instead, use a staged approach: a pilot group first, then an early adopter cohort, followed by a gradual rollout to the full population. Each phase includes structured feedback loops. What is working? What is confusing? Where are the rough edges? Early users become partners in refinement, and their input makes the broader rollout smoother. This approach also manages change fatigue; incremental deployment gives the organization time to adjust rather than overwhelming it with transformation all at once.

Monitoring and support infrastructure must be operational from the start. How will performance be tracked? Who receives alerts when something goes wrong? What escalation paths exist when users encounter problems? Establishing this infrastructure before deployment means issues get caught and resolved quickly, building confidence rather than eroding it. Too many organizations treat monitoring as something to implement "eventually," only to find themselves firefighting problems they cannot diagnose because visibility was never built in.

Communication during rollout is constant and multi-directional. Regular updates keep stakeholders informed of progress, challenges, and adjustments. Feedback mechanisms give users voice and ensure their experiences shape ongoing refinement. When problems occur, transparent communication about what happened and how it is being addressed maintains trust. Silence breeds rumor and anxiety; visibility builds confidence even when the news is not entirely positive.

Post-deployment, the transition from deployment project to operational capability requires explicit attention. Who owns the system now? What are their responsibilities? How is performance monitored over time? When does retraining occur? What budget supports ongoing maintenance? These questions must be answered before deployment teams move on to their next project. Too many AI initiatives succeed initially only to degrade because no one was assigned responsibility for their care and feeding.

Sustainment planning integrates the new capability into the organization's operational rhythm. This includes budget allocations for continued operations, staffing plans that provide the necessary expertise, and governance processes that oversee performance and address issues. The goal is to make the AI capability unremarkable, part of how work gets done rather than a special project requiring extraordinary attention.

Iteration remains essential even after full deployment. No initial rollout is perfect. User feedback, operational data, and changing requirements will drive improvements. Establish regular review cycles, quarterly or semi-annually, to assess performance, identify opportunities for improvement, and make adjustments. This continuous improvement approach keeps capabilities relevant and prevents them from becoming legacy systems that resist change.

Throughout the rollout, managing expectations is critical. Communicate realistic timelines and be honest about limitations. When delays occur, explain why and what is being done to get back on track. When capabilities do not meet initial hopes, acknowledge it and show what lessons are being applied. Leaders who overpromise and underdeliver quickly lose credibility; those who set realistic expectations and then meet or exceed them build trust that carries through setbacks.

The functional rollout mechanics also include explicit decision rights. Who can approve changes to the system? Who has the authority to pause deployment if issues emerge? Who decides when to proceed from one phase to the next? Ambiguity here creates either analysis paralysis, where no one will make decisions, or chaos, where everyone assumes they have authority. Clear governance from the start prevents both.

Resource allocation must be realistic. Rollouts require not only technical talent but also project management, change management, communication expertise, and executive attention. Budget for these resources explicitly rather than assuming they will materialize from existing capacity. Underfunded rollouts either fail or succeed by burning out the people who carry them through force of will, and burnout is not a sustainable strategy.

Finally, celebrate milestones without declaring premature victory. When a phase completes successfully, acknowledge the achievement and the people who made it possible. But remain clear that this is progress toward a goal, not the goal itself. Early celebration can create complacency that undermines subsequent phases. The balance is to maintain momentum through recognition while keeping focus on the work still ahead.

From Scattered Experiments to Strategic Portfolio

AI initiatives often begin as isolated experiments scattered across departments. A data science team in one division builds a predictive model; another group launches a chatbot pilot; a third explores computer vision for quality control. Each effort may have merit, but without a portfolio view, leaders cannot see the balance, the redundancies, or the overall value these efforts create. Portfolio management transforms experimentation into enterprise strategy.

The shift in perspective is fundamental. Each initiative must be understood as an investment, with costs, risks, and potential returns. Some are high-risk bets with transformative potential; others are steady, incremental gains that sustain momentum and build credibility. The CDAO's role is to deliberately balance these forces, constructing a portfolio that mirrors the logic of diversified investment.

A healthy portfolio includes low-risk process automation that delivers predictable gains, medium-risk predictive tools that expand existing capabilities into new domains, and a smaller allocation to high-risk innovations that could redefine operations entirely if they succeed. Too often, portfolios lean heavily toward one end of this spectrum. Some organizations play it safe, investing only in incremental tools and missing opportunities for breakthrough value. Others pursue bold, high-risk experiments without ensuring that reliable, smaller wins sustain momentum and deliver near-term credibility. A disciplined portfolio approach ensures that neither extreme dominates.

Building this portfolio begins with visibility. Create a centralized inventory of every AI initiative: pilot, proof of concept, scaled deployment, or proposal. Record each one's purpose, owner, maturity, budget, expected benefits, and risks. This visibility exposes overlap and fragmentation while revealing where resources are stretched too thin. It transforms advocacy for funding from a contest of persuasion into an evidence-based conversation.

Tracking progress across initiatives requires more than a list of projects; it demands a living view of their maturity and momentum. Each effort, whether an idea, pilot, validation, scaled deployment, or sustainment phase, should be clearly defined and regularly updated. Using consistent success and failure criteria creates discipline and transparency. When pilots underperform, retire them formally and capture the lessons learned. When they succeed, transition them with sustainment plans already in place. This rhythm of assessment and renewal prevents the quiet accumulation of half-finished work that drains both attention and resources.

Managing the portfolio also means facing hard choices. Budgets and talent are finite, and a portfolio view makes trade-offs visible. It shows where expertise is stretched too thin, where external support is essential, and where promising work must pause until capacity expands. Many organizations adopt frameworks such as stage-gate reviews or readiness-level scoring to consistently evaluate maturity and risk. These structures need not be rigid; CDAOs can adapt them to their organization's culture, by mission area, capability type, or financial priority. But the standards must remain uniform, or the discipline collapses.

A well-structured portfolio brings credibility. Systematic tracking enables leaders to demonstrate value in concrete terms: cost savings, improved accuracy, faster decisions, and mission readiness. In regulated environments, this accountability is more than a performance metric; it is a form of stewardship. Aggregated measures such as time-to-value or pilot-to-scale ratios provide a clear story of progress that executives and oversight bodies can trust.

In the end, portfolio discipline transforms the CDAO's role from project sponsor to strategist. It connects experimentation to enterprise direction and ensures that every initiative, successful or not, contributes to a coherent, cumulative whole rather than a scatter of disconnected trials.

QUICK NUMBERS
What AI Portfolios Look Like in Practice

- **60–70 percent** of AI pilots never advance beyond proof-of-concept, underscoring the need for balance and clear criteria (*"88% of AI Pilots Fail to Reach Production—But That's Not All on IT," CIO,* 25 Mar. 2025[61]).
- **6-12 months** is the typical time-to-value for scaled AI projects when portfolio frameworks are established (*Gidwani and Bhattarai, "7–Step Stage Gating Framework," IBM Think Insights,* 10 July 2025[62]).
- **80 percent** of leading organizations use some form of stage-gate or readiness-level framework to enforce decision points on scaling or retirement (*IBM Think Insights,* 2025[63]).

61. Pyle, Josh Fruhlinger. "88% of AI Pilots Fail to Reach Production—But That's Not All on IT." CIO, 25 Mar. 2025, www.cio.com/article/3850763/88-of-ai-pilots-fail-to-reach-production-but-thats-not-all-on-it.html
62. Gidwani, Priya, and Suresh Bhattarai. "7-Step Stage Gating Framework for Enterprise AI Maturity." IBM Think Insights, 10 July 2025, www.ibm.com/blog/7-step-stage-gating-framework-enterprise-ai
63. IBM Think Insights. "AI Portfolio Management Benchmarks: Balancing Risk and Return." IBM Think Insights, 2025, www.ibm.com/think/insights/ai-portfolio-management

The Real Meaning of Failing Fast

In several chapters, we have discussed the need to start small, pilot, or otherwise experiment to build momentum and credibility. Historically, a favorite phrase associated with this mentality is "fail fast". This "fail fast" motto is often recited as a slogan or published as a buzzword in a slide deck, but it's rarely practiced with integrity. The phrase has become shorthand for innovation culture, appearing in strategy documents and leadership presentations with ritualistic frequency. Yet despite its ubiquity, organizations struggle to implement it effectively. The gap between aspiration and practice reveals something important: failing fast is not fundamentally a technical challenge. It is a human one. And until leaders understand the psychological, political, and cultural forces that prevent healthy failure, the concept will remain more aspiration than reality.

The theory is compelling. In environments of uncertainty, where the path to success is unclear, the fastest way to learn is through rapid experimentation followed by honest evaluation. Projects that show early signs of failure should be terminated quickly, freeing resources for more promising work. Each failure generates knowledge that informs future decisions, creating a learning flywheel that accelerates innovation. Organizations that fail fast waste less, learn more, and move faster than those that cling to failing initiatives out of stubbornness or denial.

The practice, however, collides with human nature and organizational dynamics that make quick termination extraordinarily difficult. Consider the forces that keep zombie projects alive long after their viability has evaporated. There is the sunk cost fallacy, the psychological trap that makes teams reluctant to abandon work simply because resources have already been invested. "We've already spent six months on this" becomes justification for spending six more, even when evidence suggests the effort will never succeed. There is a career risk; the person who championed a pilot may fear that terminating it will be read as personal failure rather than organizational learning. There is political capital; leaders who pushed for funding may be reluctant to admit publicly that the bet did not pay off. There is institutional momentum; once a project is approved, budgeted, and staffed, stopping it requires active intervention rather than passive continuation, and organizations tend toward inertia.

Most insidiously, there is the optimism bias that afflicts every team close to a struggling project. When you have invested time, energy, and reputation into an initiative, you naturally interpret ambiguous signals favorably. A small improvement becomes evidence that success is just around the corner. A single positive user comment outweighs systematic feedback showing low adoption. Teams genuinely believe that one more iteration, one more tweak, one more month will turn things around. This optimism is not dishonest; it is human. But it distorts judgment and prolongs the life of initiatives that should end.

The result is the phenomenon every experienced leader recognizes: the graveyard of zombie projects, initiatives that are not quite dead and not quite alive, consuming attention without delivering value. These projects drain organizational energy, occupy talent that could be applied elsewhere, and create cynicism among staff who see resources devoted to efforts everyone privately knows will fail. Worse, they send a cultural message: in this organization, we do not actually make hard decisions. We let things drift.

The conventional response to this problem, exhortations to "just be more rigorous" or "follow the data," misses the point. The issue is not a lack of data or analytical capability. Most failing projects generate plenty of evidence of their struggles. The issue is that ending projects requires navigating human emotions, organizational politics, and cultural norms that cannot be resolved by better metrics. Any approach to failing fast that ignores these realities is doomed to remain theoretical.

So how do organizations weave in the valuable elements of rapid learning and disciplined termination without falling prey to the pitfalls that have undermined past attempts? The answer lies in designing systems that acknowledge human nature rather than pretending it does not exist.

Separate the decision from the person. The most effective organizations depersonalize failure by establishing independent review gates that are not tied to project champions. When a pilot reaches a predetermined checkpoint, a cross-functional review panel, not the team running the project, evaluates whether it should continue, pivot, or terminate. This structure creates psychological distance between the decision and the individuals involved. The team can present their best case for continuation, but the decision rests with people whose careers are not tied to the project's survival. This does not eliminate emotional investment, but it reduces the correlation between project continuation and personal reputation.

Normalize closure as success. One of the most powerful cultural interventions is redefining what counts as a successful pilot. In traditional cultures, success means scaling; anything short of that is failure. This binary framing makes early termination painful and rare. More sophisticated organizations recognize that generating reliable knowledge is success, regardless of whether the pilot scales. A pilot that proves a hypothesis false has succeeded in producing valuable information. A pilot that reveals unexpected data quality issues has successfully surfaced a problem that might have undermined future work. When leaders celebrate these outcomes as wins, teams become more willing to share negative results and recommend termination when appropriate.

Build termination into the timeline. Pilots should have predetermined end dates, not open-ended timelines that drift based on results. This temporal boundary creates a natural decision point where continuation requires active justification rather than passive drift. At that point, the question is not "Should we kill this?" which implies failure, but "Has this earned continuation?" which shifts the burden of proof. This framing makes termination the default unless evidence argues otherwise, countering the organizational inertia that keeps marginal projects alive.

Harvest value from every ending. When a pilot terminates, treat it as a reusable asset rather than a write-off. Conduct structured post-mortems that separate what was learned about the technology, the data, the organization, and the process. Document these insights in formats that other teams can access: a lessons-learned repository, a data quality registry, a model performance database. Circulate findings across the organization so that what one pilot learned about bias detection or user adoption or integration challenges becomes institutional knowledge. This practice transforms termination from loss into contribution. The team that ended early did not fail; they generated knowledge that will prevent others from repeating the same mistakes.

Create safe language for struggle. The vocabulary organizations use to discuss challenging projects matters enormously. "Failure" carries connotations of incompetence and blame. "Learning" sounds evasive, as if trying to avoid accountability. More effective language acknowledges both the outcome and the value: "This pilot did not validate our hypothesis, but it revealed critical constraints about our data infrastructure that will inform future work." This framing is honest about what did not work while recognizing what was gained. When leaders model this language consistently, teams internalize permission to report problems without catastrophizing them.

Address the sunk cost fallacy explicitly. Sunk costs are a cognitive trap, and awareness alone does not eliminate them. Organizations that fail fast well include explicit prompts in review processes that force teams to separate past investment from future potential. Review templates might include questions like: "If we were starting today with zero investment, would we pursue this approach?" or "What would have to be true for this to succeed, and how likely is that?" These questions redirect attention from what has been spent to what could be gained going forward, enabling more rational decisions.

Distinguish between project failure and execution failure. Not all unsuccessful pilots are created equal. Some fail because the hypothesis was wrong; these are good failures that generated knowledge quickly. Some fail because of poor execution: inadequate data preparation, unclear requirements, or insufficient stakeholder engagement. These are bad failures that wasted resources without producing reliable insights. Organizations must be able to distinguish between them. The former should be celebrated and harvested for learning. The latter should trigger process improvements and, if patterns emerge, accountability conversations. Conflating the two creates confusion about what behaviors to reinforce and what to correct.

Manage public narratives carefully. How failure is communicated to the broader organization or external stakeholders affects whether teams will be honest about struggles in the future. If terminating a pilot is framed as "we wasted money and failed," teams will hide problems to avoid that narrative. If it is framed as "we tested an important hypothesis and learned it does not apply in our context, saving us from a much larger investment later," termination becomes defensible. Leaders must actively control this narrative, ensuring public communication reinforces the cultural norms they want to establish.

Protect the people, not the project. One reason teams resist terminating pilots is fear that the people involved will be penalized: reassigned to less desirable work, passed over for promotions, or stigmatized as having been part of a failure. Organizations that fail fast successfully make clear that individuals will not suffer career consequences for projects that end early when they are executed well and communicated honestly. Protecting people while ending projects requires deliberate action: publicly recognizing the team's contributions, ensuring they move to roles of equal or greater opportunity, and including their work in performance evaluations as valuable learning rather than failure. When staff see that colleagues who terminated projects are treated well, fear diminishes.

Balance rigor with empathy. Ending a pilot is often emotionally difficult for the people involved, even when they intellectually agree it is the right decision. They have invested time, creativity, and effort. They may have grown attached to the vision of what the project could become. Leaders who dismiss these emotions as unprofessional or irrelevant create resentment and resistance. More effective leaders acknowledge the difficulty: "I know this is disappointing. You put real work into this, and it is hard to let go. But the evidence is clear, and we owe it to the organization to redirect these resources." This empathy does not change the decision, but it changes how people experience it, making future honesty more likely.

Track the whole portfolio, not just individual projects. When every project is evaluated in isolation, teams feel that each one must succeed to justify their work. Portfolio-level tracking changes this dynamic. If leadership evaluates a team based on the overall value generated across multiple experiments, with the understanding that some will not pan out, individual project termination becomes less threatening. The question becomes not "Did this project succeed?" but "Is this team learning quickly and deploying resources effectively?" Portfolio thinking creates permission for individual experiments to fail as long as the aggregate delivers value.

Implementing these practices requires commitment from the top. If senior leaders publicly celebrate failed pilots that generated learning, middle managers will follow suit. If executives ask hard questions about struggling projects and reward honest answers, teams will be more forthcoming. If the CDAO models the language of disciplined learning rather than blame, that vocabulary will spread.

But even with perfect process and cultural support, some projects will still linger too long. Human attachment to work cannot be entirely engineered away, nor should it be. The goal is not to eliminate all friction around ending projects, but to reduce it enough that termination happens more readily when warranted. Organizations that fail fast do not eliminate the difficulty; they make it manageable.

Real agility means having the courage to stop when evidence says a project will not scale, even when stopping is uncomfortable. It means acknowledging that speed is not the same as success, and that pushing pilots forward long after the data turns negative is not persistence but denial. It means treating each experiment as contributing to a larger portfolio, not as an isolated project whose failure would reflect poorly on its sponsors. When successful pilots scale, the infrastructure for transparency and auditability should already be in place. A brilliant model that cannot be explained or reproduced is not a success; it is a liability waiting to be discovered.

Failing fast, done well, is not reckless. It is disciplined learning. A clear hypothesis, measurable thresholds, and transparent review gates allow teams to pivot or shut down with integrity. This protects budgets and preserves morale. It also signals maturity: leaders who know when to quit gain credibility with executives and oversight bodies alike. When a pilot ends early, it should be treated as a reusable asset. Conduct brief post-mortems, document technical lessons, capture process insights, and circulate findings across teams. What one pilot learned about data quality or model drift can prevent waste elsewhere.

When organizations normalize this transparency, fear diminishes and experimentation thrives. A culture that fails fast responsibly replaces blame with evidence. It turns mistakes into material for progress. But getting there requires more than slogans. It requires designing systems that work with human nature rather than against it, creating structures that make healthy failure possible even when it remains difficult. The organizations that succeed at this do not find it easy. They simply find it worth the effort.

VOICES IN THE FIELD
What Leaders Say About Failing Fast

"We do not celebrate failure for its own sake. What we celebrate is the discipline to stop a project when the evidence tells us it will not scale. Every early ending is money we saved and time we redirected to something that works."
–Chief Data Officer, Federal Agency

"Our pilots are designed with exit ramps. If we are not hitting benchmarks within months, we shut it down, but we never walk away empty-handed. We harvest every learning, document every misstep, and make the next pilot smarter."
–VP of Analytics, Healthcare Organization

"In AI development, the worst outcome is not failure, it is ambiguity. If you cannot say whether something succeeded or failed, you cannot move forward."
–Chief AI Scientist, Defense Contractor

"We call it 'failing forward.' Even when a model does not work, it teaches us about our data, our processes, or our culture."
–Head of Digital Transformation, Financial Services

"Ending projects early has actually increased trust with leadership. They see that we are disciplined. We know when to double down and when to walk away."
–Program Manager, Global Supply Chain

Measuring What Matters

As you move through these pilots, measurement becomes central to support, transparency, and momentum. Measurement transforms ambition into credibility. Without metrics, success is opinion; with them, it becomes evidence. Yet measuring AI value is notoriously difficult. Technical performance rarely maps cleanly to business outcomes, and the temptation to claim victory based on model accuracy alone can obscure whether anyone is actually using the tool or whether it is improving decisions.

The discipline begins before any pilot launches: define baselines. If the goal is faster case resolution, record the current average time. If the objective is fraud detection, capture the existing false-positive rates. Baselines turn the abstract notion of "improvement" into quantifiable progress that leadership can understand and trust.

Technical metrics matter, but they must be linked explicitly to mission outcomes. A system may score high on precision but still fail operationally if adoption is low or if it merely shifts workload from one team to another without net benefit. Leaders should pair hard metrics like cost savings and accuracy improvements with softer indicators such as user trust, process insights, and the quality of downstream decisions. The combination paints a more accurate picture of value than either category alone.

Equally important is time-to-first-value: the period from pilot launch to measurable outcome. Shortening that cycle accelerates confidence and momentum. Organizations that can demonstrate early wins, even modest ones, build the political capital needed to sustain investment through the inevitable setbacks.

Adoption is often the missing link. Even the most accurate model delivers no benefit if it sits unused. Usage rates, integration depth, and downstream decision changes are leading indicators of realized value. A fraud detection model with perfect precision means nothing if investigators ignore its alerts. A demand forecasting tool achieves little if planners override its recommendations without understanding why.

Measuring AI requires both data science and storytelling: the ability to show not only what the model achieved technically, but also what that achievement means for the mission. The CDAO who can translate a reduction in false positives into hours saved for investigators, or a faster inference time into lives protected, will always have an easier time securing resources than one who speaks only in technical abstractions.

QUICK NUMBERS
Measuring AI Value in Practice

- **60–70 percent** of organizations struggle to link AI results to business outcomes (*TechRadar*, 8 July 2025[64]).
- **25–30 percent ROI** within the first year is a standard benchmark in finance and logistics (*Agility–at–Scale*, 2025[65]).
- **1 in 3** technically successful systems fail to deliver value because user adoption stays below 50 percent (*TechRadar*, 2025[66]).

Building for Durability

Every AI system carries risk. Models drift as the world changes beneath them. Data sources degrade or introduce unexpected bias. Users find ways to misuse tools their designers never anticipated. External disruptions, from regulatory changes to adversarial attacks, can render yesterday's solution tomorrow's liability. Ignoring these realities invites crisis. Responsible teams plan for failure the way pilots plan for turbulence: not because they expect disaster, but because preparation is what makes survival routine.

Risks fall into three broad classes. Data risk stems from degraded or biased inputs; regular sampling and validation guard against drift. Model risk includes overfitting and the accumulation of technical debt; version control and scheduled retraining reduce exposure. Human risk, including misuse, misunderstanding, or overreliance on automated outputs, is often the most dangerous and the hardest to anticipate. Training programs and clear escalation channels mitigate it, but vigilance must be continuous.

64. "State of AI Value Realization 2025: Metrics that Matter." TechRadar Pro, 8 July 2025, www.techradar.com/pro/state-of-ai-value-realization-2025-metrics-that-matter
65. "Measuring Enterprise AI ROI: A Data-Driven Framework." Agility-at-Scale, 2025, www.agilityatscale.com/reports/measuring-enterprise-ai-roi
66. "State of AI Value Realization 2025: Metrics that Matter." TechRadar Pro, 8 July 2025, www.techradar.com/pro/state-of-ai-value-realization-2025-metrics-that-matter

Contingency planning starts early, not after deployment. Define rollback procedures before launch. Set performance thresholds that trigger human review or model suspension. Establish alternate workflows for critical decisions so that a failing system does not bring operations to a halt. Transparent documentation supports accountability. When regulators or oversight bodies ask how an error occurred, the ability to trace data lineage and model versions can be the difference between a manageable correction and an organizational crisis.

Risk management is not about eliminating failure; it is about catching failure safely. The most advanced organizations treat resilience as a design requirement, not an afterthought.

AI Lifecycle and Sustainment Planning

AI models are living systems. Their relevance fades unless maintained. Sustainment planning ensures that success today does not become tomorrow's technical debt.

Lifecycle management spans monitoring, retraining, governance, and resource allocation. Continuous monitoring detects drift before it causes harm. Scheduled retraining keeps models aligned with evolving reality. Governance provides the guardrails that make this maintenance auditable and defensible. None of this happens automatically; it requires deliberate investment.

Budgeting for sustainment is essential, yet frequently neglected. Funding only for development leaves teams scrambling later, forced to choose between letting models decay or diverting resources from new initiatives. The most effective CDAOs treat AI like infrastructure, with operational budgets for upkeep rather than just capital investment for launch. Communicate this plainly to executives: sustainment is not a cost overrun. It is the cost of reliability.

QUICK NUMBERS
Lifecycle in Motion

- **1 in 4** scaled AI systems require significant retraining within 24 months (*IBM Think Insights*, 2025[67]).
- **70 percent** of failed AI programs cite neglected maintenance as a core cause (*McKinsey & Company*, 2024[68]).

The Human Architecture of AI

No AI strategy works without people. Talent planning covers both the specialists who build and maintain systems and the broader workforce that interprets outputs and applies them to real decisions. Neglect either group, and even the most sophisticated technology will fail to deliver value.

A balanced approach builds a core team of data scientists, engineers, and architects while simultaneously empowering adjacent talent: analysts, domain experts, and business leads who may never write code but whose judgment determines whether AI insights translate into action. Training these partners in data literacy yields faster, more sustainable adoption than any technical improvement alone. When a nurse understands what a predictive model can and cannot tell her, she uses it wisely. When she does not, she either ignores it entirely or follows it blindly, and both paths lead to harm.

Upskilling must be continuous, not a one-time event. Microlearning modules, embedded exercises, and scenario simulations help non-technical staff understand how to question AI responsibly. The goal is not to turn everyone into a data scientist, but to create a baseline of critical thinking that allows the organization to use these tools without abdicating human judgment.

External partners can enhance capability, but relationships must be structured to transfer knowledge rather than create dependence. Contracts should include co-development requirements and documentation

67. IBM Think Insights. "AI Portfolio Management Benchmarks: Balancing Risk and Return." IBM Think Insights, 2025, www.ibm.com/think/insights/ai-portfolio-management
68. "The State of AI in 2024: From Experimentation to Implementation." McKinsey & Company, 2024, www.mckinsey.com/featured-insights/artificial-intelligence/state-of-ai-2024

standards that ensure continuity when vendors leave. An organization that cannot sustain a system without its original builders has not truly succeeded; it has merely rented capability.

Retention hinges on purpose and growth. Specialists stay where they see impact and opportunity, where their work connects to outcomes that matter. Broader staff remain engaged when they feel AI enhances rather than threatens their work. Talent strategy is not a staffing chart; it is the human architecture of transformation.

VOICES IN THE FIELD
Building AI Talent and Literacy

"Our success wasn't about hiring PhDs, it was about helping nurses and administrators understand what the tools were telling them."
–Chief Analytics Officer, Healthcare System

"Every vendor contract includes joint teams. If our staff can't sustain a system after the vendor leaves, we haven't succeeded."
–CDAO, Federal Agency

"Our engineers sit beside domain experts. Breakthroughs come not from better algorithms, but from aligning models with mission reality."
–Director of AI Programs, Defense Contractor

Ethics as Engineering

AI and ethics are a complicated landscape that is constantly evolving. There are volumes of content on this subject, so it didn't make sense to go too deep. That said, it's worth noting that ethics and responsibility are not obstacles to innovation; they are its stabilizers. Organizations that embed responsible practices early save time, reduce costs, and protect their reputations later. Bias, opacity, and misuse are predictable if left unchecked. Preventing them requires integration, not bureaucracy.

The work begins during pilot design, when teams review data sources for potential bias before a single model is trained. It continues during validation, when subgroup performance is measured rather than just overall accuracy. It persists through deployment, when audit trails capture model decisions for later examination. These steps fold naturally into existing review gates; they do not require separate ethics committees or parallel approval processes, though such structures may be valuable for high-stakes applications.

Transparency and explainability sustain trust. Users need to understand why a model recommends an action, even if they cannot follow the mathematics. Simple model cards or plain-language summaries often suffice. The goal is not to make every user a machine learning expert, but to ensure that no one is asked to follow a recommendation they cannot interrogate.

Human oversight remains essential. Automation should accelerate judgment, not replace it. For high-stakes decisions, escalation channels must be in place for uncertain cases, protecting both the system's integrity and the people affected by its outputs. When responsible practice is embedded well, it becomes invisible. It simply defines how AI is built and used, unremarkable because it is universal.

Making Progress Visible

Even sound strategies falter if misunderstood. Communication is how leaders turn progress into shared belief across the organization.

Excitement should come from outcomes, not hype. Share tangible wins: a reduction in case backlogs, faster forecasting cycles, and improved safety metrics. Celebrate collaboration rather than technology for its own sake. When a cross-functional team delivers results, make their partnership as visible as the tools they used.

Address fear openly and honestly. Show how AI augments rather than replaces human work. Explicitly link initiatives to organizational values: efficiency, fairness, security, and mission impact. People support what reflects their principles. A workforce that sees AI as aligned with what they care about will adopt it far more readily than one that views it as imposed from above.

Keep the story continuous. Communicate during pilots, through scaling, and into sustainment. Regular updates reinforce momentum and demonstrate transparency. Silence breeds suspicion; even modest progress, shared consistently, builds confidence.

Empower local champions to tell the story in their own contexts. When managers and frontline users explain AI in their own words, adoption accelerates. Strategic communication turns technical progress into organizational confidence. It ensures AI is not a mystery imposed by specialists, but a shared success owned by everyone.

AI as a Living Discipline

AI and analytics are never finished; they evolve with the organizations that use them. For the Chief Data and AI Officer, this is both a challenge and an opportunity. No single model or platform defines success; what matters is the discipline that endures beyond any project.

That discipline begins with clarity about current capabilities. It advances through structured experimentation, treating every pilot as an opportunity to learn. It matures through portfolio oversight that connects individual initiatives to enterprise direction. And it sustains itself through governance, talent development, and ethical practice, making responsible AI the default rather than the exception.

The work of the CDAO is not to chase hype but to build durability: to shape an organization that adapts, learns, and thrives in complexity. When leaders treat AI as a living discipline, they build more than systems. They build capacity for perpetual improvement, a foundation that will serve the enterprise long after today's models are obsolete.

AI's future will belong not to those who rush, but to those who build steadily, responsibly, and together.

Chapter 8

METRICS & MISSION INTEGRATION

Every organization knows the feeling of investing in data or technology programs without being able to clearly state what those investments accomplished. The language of return on investment has always been difficult for leaders in this space, not because results are absent, but because they are too often buried beneath layers of technical detail or expressed in terms that do not connect to the outcomes people care about most. When a new system is deployed, a data pipeline is automated, or a machine learning model is trained, the benefits are rarely captured in ways that resonate with decision makers or those closest to the mission. This has left many programs vulnerable to shifting budgets, skeptical stakeholders, and a persistent sense that progress is difficult to measure.

This chapter is about resolving that tension. The work of a Chief Data and AI Officer, or any leader responsible for organizational transformation, is not just about delivering technology. It is also about creating a visible, sustained impact tied to the objectives at the heart of the enterprise. Measurement, when done with care and foresight, becomes a bridge between technical achievement and organizational value. It shows that the time and resources spent on data and AI are not abstract investments but concrete steps toward stronger readiness, better services, and improved efficiency.

Previous chapters have explored the importance of aligning data and AI initiatives with strategy, of building resilient organizations, and of designing programs that endure. Here, the focus shifts to how success is captured and communicated once those initiatives are underway. Alignment without measurement risks becoming performative. Modernization without proof of impact risks being dismissed as another passing trend. What separates meaningful programs from fleeting ones is the ability to quantify their effect on outcomes that matter to leadership, operators, and stakeholders alike.

The emphasis in what follows is not simply on numbers or dashboards, but on the stories they enable leaders to tell and the confidence they build across the organization. When metrics illuminate impact in a way that leadership can see and trust, the result is not just validation of past work, but a stronger foundation for future investment.

The Translation Imperative

For every technical breakthrough within an organization, there is a moment when leadership asks, "So what does this mean for us?" The answer to that question is often where data leaders falter. A model might achieve higher precision than anything used before. A new data pipeline might dramatically reduce processing time. An analytic dashboard might finally bring together insights scattered across multiple systems. These are accomplishments worth celebrating, but unless they are directly connected to mission priorities, their value remains locked within technical circles, invisible to the very people whose support determines whether the work continues.

The operational challenge is straightforward to describe but difficult to execute. Most leaders in an organization do not think in terms of model accuracy or data lineage. They think in terms of readiness rates, customer satisfaction, risk mitigation, cost efficiency, or delivery speed. The CDAO's role is to ensure that progress in the technical domain is explained in these terms. A predictive maintenance tool, for example, is not just a demonstration of algorithmic sophistication. It is a system that increases aircraft availability, reduces mission delays, and allows skilled maintainers to focus on the most critical tasks. Natural language processing is not just about tokenization or entity recognition. It is a tool that helps legal teams process cases faster, reducing backlog and improving outcomes for clients or citizens.

This translation must begin before a project starts, not after it delivers results. Technical objectives should be mapped directly to mission outcomes from the outset, with that mapping documented and socialized with stakeholders so that all parties are aligned on what success looks like in terms of impact. No data initiative should exist in isolation. Every investment should be clearly tied to a measurable objective that leaders already care about. When this discipline is absent, teams scramble after the fact to explain why their work mattered, often failing to make the case convincingly because the connection was never established in the first place.

Translation also requires mechanisms for regular communication across technical and mission teams. This goes beyond periodic briefings. It involves creating shared artifacts: mission-focused dashboards that operators actually use, annotated workflows that show where data and AI are improving decision points, and joint working sessions where technical staff and operators review results together. These mechanisms help both sides see how incremental improvements in the technical space lead to outcomes in the operational space. They also build relationships that make future collaboration easier, because trust has been established through shared work rather than through formal reporting alone.

Data leaders must also become skilled storytellers who connect innovation to value. This does not mean oversimplifying or stripping away technical detail. It means layering the story so that different audiences can engage at the right depth. A commander or executive may only need to know that a new analytic reduces decision time by thirty percent. A technical peer may want to know how that outcome was achieved. Both perspectives must be honored, but it is the data leader's responsibility to create coherence between them. The executive who hears only technical jargon will tune out; the engineer who hears only simplified summaries will lose confidence that leadership understands the work. Skilled translation serves both.

Sustaining this translation requires constant validation. It is not enough to connect technical progress to mission value at the start of a project. Conditions change, missions evolve, and what counted as success six months ago may no longer be relevant today. The CDAO and their team must regularly revisit and recalibrate the mapping between technical and mission outcomes. This discipline is what keeps the organization focused on delivering value that is both current and enduring.

Finally, data leaders should cultivate a posture of partnership rather than service. Too many technical teams position themselves as providers of capability to mission stakeholders, waiting for requirements to arrive and then delivering solutions that meet the stated specifications. The more effective model is to operate as collaborators, sharing accountability for mission success. When the data team owns not only the technical deliverable but also its translation into measurable mission impact, they move from being a support function to being a strategic partner. This shift in posture strengthens trust, accelerates adoption, and makes it far more likely that data and AI initiatives will endure beyond their initial champions.

The heart of this work is translation, but translation in the operational sense is more than words. It is about designing measurement frameworks, communication channels, and partnership models that keep value visible at every stage of the journey. It ensures that data and AI do not remain behind the curtain, but instead step onto the stage as visible, measurable drivers of the outcomes that matter most.

From Pilot to Permanence

Many organizations know how to launch a promising pilot. Far fewer know how to sustain that promise over time. Pilots are designed to prove that a concept can work, but they are not designed to endure. Too often, they remain in a demonstration state, celebrated briefly before being overtaken by other priorities or left unsupported when the initial funding cycle ends. The result is an innovation portfolio that feels impressive on paper yet leaves little lasting change on the ground. Conference presentations accumulate while operational impact remains elusive.

This pattern is not unique to data and AI. Technology modernization efforts have often failed when they remained trapped in demonstration mode, never making the difficult transition to operational reality. The same risk applies here. A pilot that is not fully embedded into mission workflows, resourced with a plan for sustainment, and measured against outcomes that leaders care about will remain an isolated success rather than a durable transformation. Sustaining impact beyond the pilot is therefore one of the most critical responsibilities of a Chief Data and AI Officer, and one of the least glamorous.

The work begins with integration. A capability must move from being "something the data team built" to being "something the mission relies on." This requires deliberate effort to embed new processes, tools, and analytic outputs into the organization's standard operating environment. If a logistics AI model predicts supply shortages, its outputs should not be viewed in isolation on a separate dashboard. They must feed into the same systems and decision cycles that logisticians and commanders already use to plan operations. Only then will the tool shift from being a pilot to being part of the organization's muscle memory, something people use without thinking about where it came from.

Sustainment requires alignment of resources, authorities, and accountability from the very beginning. This means budgeting for operations and maintenance from the outset, not as an afterthought once the pilot is complete. It means identifying who will own the system after initial implementation and ensuring they have the training, staffing, and authority to continue driving impact. Too many pilots fail because ownership remains ambiguous, with no clear transition from innovation teams to operational teams. The practitioner can address this by naming the eventual owner early, engaging them throughout development, and building their responsibilities directly into the sustainment plan. When the owner is surprised to learn they are responsible, the pilot has already begun to fail.

Measurement plays a decisive role in sustaining momentum. When leaders see the impact of a new tool reflected in the same outcome reports they already rely upon, confidence grows. Sustained reporting that highlights improvements in readiness, cost savings, or decision speed provides proof that the initiative is not a one-time experiment but an enduring contributor to mission success. This is why the mapping between technical metrics and mission outcomes, established during the translation phase, must be carried forward into sustainment. The metrics that justified the pilot must continue to demonstrate value long after the initial excitement has faded.

Cultural adoption is equally critical. Sustaining impact requires more than system integration and resource allocation. It requires people across the enterprise to recognize the new capability as part of their day-to-day work. This involves training, change management, and continuous engagement with those who use the capability daily. Success is achieved when users no longer think of the new system as separate or novel. Instead, it becomes simply a matter of how work is done. The tool disappears into the workflow, which is the highest form of praise an innovation can receive.

Practitioners should also anticipate the risk of drift. Even when a pilot is initially successful, changing operational conditions can render it less relevant. A model trained on one set of patterns may degrade as the environment evolves. A dashboard that provides critical insights today may no longer capture tomorrow's priorities. To guard against this, sustainment must include processes for ongoing validation and recalibration. Feedback loops should allow operators to flag when outputs no longer reflect their reality, and data teams should have the resources to retrain or reconfigure solutions accordingly. Without this vigilance, a successful pilot can quietly become a neglected liability.

Finally, leaders must communicate sustained impact as clearly as they communicated early wins. Early pilots often generate enthusiasm because they are new. Sustained capabilities risk fading into the background precisely because they become normal. The CDAO's job is to remind leadership and stakeholders that normal is the goal. What was once an experiment has now become a reliable part of mission execution. Regularly communicating that success helps ensure continued resourcing and builds organizational trust in future data and AI initiatives. Silence, by contrast, invites the assumption that nothing is happening, which makes the next budget cycle considerably harder.

Sustaining impact beyond the pilot is not glamorous work. It requires planning, patience, and constant reinforcement. Yet it is the difference between innovation theater and transformation. When data and AI initiatives are integrated into mission workflows, supported by clear ownership, tied to measurable outcomes, and embraced by users, they move beyond the narrow boundaries of a pilot and become enduring assets. This is where real progress happens, and where the value of a data and AI strategy is truly realized.

TREND SNAPSHOT
Why Pilots Stall and What It Takes to Stick

- **Transformations still miss the mark.** Large programs fail at a stubborn rate. McKinsey notes that roughly 70 percent of corporate transformations do not achieve their goals. (McKinsey 2019[69])

- **AI pilots are being abandoned at higher rates.** S&P Global's 2025 survey of 1,006 leaders found a sharp rise in projects dropped before production. Forty-two percent of companies abandoned most AI initiatives, and on average 46 percent of proofs of concept were scrapped between pilot and broad adoption. (S&P Global 2025[70])

- **Federal agencies feel the same friction.** A GDIT survey reported by Nextgov found that 58 percent of agency pilots failed to advance due to scalability, funding, adoption, or ATO hurdles. Even for successes, the average time from pilot to production was 14 months. Predictive analytics and machine learning pilots were most likely to cross the line, while generative AI lagged. (Nextgov 2024[71])

- **Only a small share sees financial lift.** MIT Sloan Management Review and BCG reported that only 10 percent of organizations achieved significant financial benefits from AI, underscoring the need to pair technology with process change and learning loops. (BCG 2020[72])

- **New research flags a GenAI divide.** Recent MIT coverage indicates that about 95 percent of generative AI pilots are failing to show measurable profit and loss impact, pointing to

69. "Why do most transformations fail? A conversation with Harry Robinson." McKinsey & Company, 10 July 2019, https://www.mckinsey.com/capabilities/transformation/our-insights/why-do-most-transformations-fail-a-conversation-with-harry-robinson
70. "AI experiences rapid adoption, but with mixed outcomes – Highlights…" S&P Global Market Intelligence, https://www.spglobal.com/market-intelligence/en/news-insights/research/ai-experiences-rapid-adoption-but-with-mixed-outcomes-highlights-from-vote-ai-machine-learning
71. Graham, Edward. "Agencies are advancing AI pilots but face scalability challenges, report finds." Nextgov, 9 May 2024, https://www.nextgov.com/emerging-tech/2024/05/agencies-are-advancing-ai-pilots-face-scalability-challenges-report-finds/396439/
72. "Annual MIT Sloan Management Review—Boston Consulting Group Study Finds Significant Financial Benefits with AI…" MIT Sloan Management Review, 20 October 2020, https://www.bcg.com/press/20october2020-study-finds-significant-financial-benefits-with-ai .

integration and workflow issues rather than model quality alone. (Computerworld 2025[73])

What this means for practitioners

1. **Plan sustainment on day one.** Define the operating owner, support model, and budget line for operations and maintenance before the pilot begins. Tie these to mission milestones that already exist in the organization's rhythm.
2. **Embed outputs in the work, not just a dashboard.** Route model signals into existing case systems, planning boards, or command rhythms so that operators consume results inside their normal flow.
3. **Gate progression with outcome metrics.** Use a simple chain for every use case: technical validity, process improvement, and mission outcome. Advance only when each link is proven with real data, and report those outcomes in the same forums leadership already uses.

The Art of Communicating Results

The success of a data or AI initiative does not speak for itself. A new capability might reduce processing time by half or improve accuracy across a critical workflow, but without clear communication, these results can remain invisible to the very leaders whose support is required for continued growth. Communicating results is not a public relations exercise. It is a discipline of leadership that connects technical progress to organizational trust and, ultimately, to sustained investment.

The first requirement is audience mapping. Different groups consume information differently, and a message crafted for one audience may fall flat with another. A senior executive may have only a few minutes in a briefing and will care most about whether a capability improves readiness, reduces costs, or strengthens compliance. A mission stakeholder will want

73. Erikson, Victor. "Study: 95 Percent of Corporate Generative AI Projects Fail." Computerworld, 2025, www.computerworld.com/article/4042361/study-95-percent-of-corporate-generative-ai-projects-fail.html .

to understand how the tool helps them deliver on daily objectives, whether that is case resolution, flight operations, or health service delivery. Technical staff may prefer to see the precision scores, architecture diagrams, or throughput statistics. The data leader's responsibility is to build a communication framework that speaks to each group at the level that matters to them, while maintaining a consistent underlying narrative. The story must be one story, told at different depths.

The second requirement is to connect results to existing reporting rhythms. Leaders already have standing reports, operational reviews, and performance dashboards to assess progress. If data and AI results are presented in separate channels, they risk being seen as additional or optional, something to check when time permits rather than something central to operations. When results are embedded into the metrics that leaders already review, the impact becomes part of the organizational bloodstream. Instead of creating a stand-alone dashboard for fraud detection improvements, for example, the CDAO can work to insert the relevant performance measures into the same reports that financial executives already use to track risk and audit performance. Visibility in existing channels is worth far more than prominence in new ones.

Third, practitioners should anchor communication in outcomes that are understandable without translation. This does not mean avoiding technical detail altogether, but it does mean that leading with jargon will undermine credibility. A more effective approach is to begin with outcomes that everyone recognizes, then show how technical progress enabled them. Rather than beginning with the details of a model's recall rate, a CDAO might open with the fact that the backlog of investigative cases dropped by twenty percent in six months. The technical explanation can follow, but the opening message makes clear that the work is already producing value. Executives remember the outcome; engineers can follow up on the method.

Storytelling is another essential dimension. Data leaders often underestimate the power of narrative in shaping how results are received. Numbers provide proof, but stories create meaning. An executive will remember that "veterans are now receiving their benefits weeks faster" more vividly than a performance chart showing workflow acceleration. This does not mean inventing anecdotes, but it does mean elevating the voices of users and stakeholders who have felt the change. Including these perspectives in communications not only validates the results but also builds advocacy across the organization. When a frontline operator explains how a tool changed their work, the message carries weight that no chart can match.

Timing matters as well. Success communicated too early can appear premature, raising expectations that subsequent setbacks will disappoint. Success communicated too late can lose relevance, arriving after decisions have already been made or attention has moved elsewhere. Practitioners should establish a cadence that provides updates at key decision points, mirroring the project lifecycle: early indicators that validate the approach, operational measures that show the tool working in practice, and mission outcomes that demonstrate lasting impact. Done well, this sequence allows leaders to see progress accumulate, building confidence that each stage is a step toward lasting transformation rather than an isolated win.

Finally, communication must be sustained, not episodic. Pilots are often celebrated with high visibility, only to fade into the background once they become routine. As discussed earlier, routine is the goal, but that does not mean silence. Practitioners should develop mechanisms to continue reporting value long after the initial launch, ensuring that results remain visible, measurable, and aligned with the organization's evolving priorities. This creates a feedback loop that reinforces trust and secures support for the next wave of innovation. The alternative is a pattern where each new initiative must fight for attention from scratch, never building on the credibility that sustained communication could provide.

Communicating results is therefore not about polishing the message, but about building a system of transparency and trust. It ensures that leaders see data and AI not as abstract technology projects, but as concrete enablers of mission success. By embedding communication into existing rhythms, layering detail to meet different audiences, and combining metrics with stories, the CDAO ensures that progress is visible, credible, and compelling. This is what turns one-time wins into sustained momentum and lays the foundation for future growth.

VOICES IN THE FIELD
Communicating Results with Impact

"The only way my leadership team paid attention was when I stopped talking about model accuracy and started talking about mission hours saved. Once they saw that our analysts could close cases faster, they began asking when we could scale the tool. That shift in framing changed everything."
–Senior Data Officer, Department of Justice

"We made the mistake early on of creating a separate AI dashboard that no one checked. Once we placed our performance metrics inside the operational readiness reports that commanders already reviewed weekly, our progress was impossible to ignore. It went from being an experiment to being part of the mission conversation."
–Program Manager, Defense Logistics Agency

"What resonated most with executives was not the chart that showed error reduction. It was the story of a single customer who received their benefit weeks earlier than expected. That single story, paired with the data, created the momentum for continued investment."
–Director of Analytics, Federal Health Agency

"Communicating results requires discipline. You cannot just wait until the end of a project. You need to map out your communication milestones at the same time you design your technical milestones. That is how you build confidence step by step."
–Chief Data Scientist, Financial Services Sector

Accountability as Architecture

Accountability is the thread that connects strategy, measurement, and mission value. Without it, even the most promising initiatives risk becoming little more than technical experiments that fade once leadership attention shifts. Building a culture of accountability is not about assigning blame when results fall short. It is about creating the structures, habits, and expectations that keep everyone aligned on outcomes and ensure that progress is both visible and sustained.

Accountability is the active layer that ensures governance structures are lived out in daily operations. It is where the CDAO moves from being a strategist to being a steward of impact. Data leaders must position accountability as a shared responsibility across technical teams, mission stakeholders, and executives. When accountability is distributed in this way, it creates resilience, because no single point of failure can derail progress. When accountability is concentrated in a single person or team, the entire edifice becomes fragile, dependent on continued attention from a narrow group.

The first step is to set clear expectations for outcomes. Too often, projects begin with enthusiasm but lack precise definitions of what success will look like and who is responsible for achieving it. Establishing accountability starts with defining measurable objectives at the outset and agreeing on who owns each objective. This agreement should not remain only in project documentation. It should be socialized and reinforced in the governance forums that oversee progress. By tying ownership to existing governance processes, accountability becomes part of the normal rhythm of organizational oversight rather than an additional burden layered on top.

The second step is to design transparent, consistent accountability mechanisms. Dashboards, regular reviews, and after-action reports all serve this function, but they must be integrated into the same channels that leadership already uses to make decisions. If accountability measures are kept in separate systems, they will be sidelined. If they are woven into existing oversight and mission reporting, they will become unavoidable. Visibility is a form of accountability, and when leaders can see how investments are performing against mission outcomes, they are better able to reinforce progress and correct course when needed.

Third, practitioners must balance accountability with empowerment. A culture of accountability fails if people feel that their only role is to defend their work against criticism. It succeeds when people feel trusted to make decisions, take informed risks, and adapt when evidence shows that a different approach is required. Leaders can foster this balance by making clear that accountability includes the freedom to pivot. If a solution is not producing the intended outcome, acknowledging the shortfall and redirecting resources should be recognized as responsible leadership, not as failure. This approach aligns with the emphasis on evidence-based iteration and the acceptance of productive failures that runs throughout this book.

Fourth, practitioners should invest in feedback loops that allow accountability to flow in multiple directions. Operators must be able to signal when systems are not meeting their needs. Data teams must be able to provide transparent updates on progress and challenges. Executives must demonstrate that they are acting on this information. These feedback loops build trust and ensure accountability is not imposed from the top down, but is shared across levels of the organization. When feedback flows freely, problems surface early enough to be addressed. When feedback is blocked, small issues become large failures.

Finally, accountability must be tied to both recognition and correction. Celebrating teams when their work delivers measurable mission value reinforces the behaviors the organization wants to sustain. Public acknowledgment of impact, whether in leadership briefings or internal communications, builds morale and motivates others to approach their work with the same commitment to outcomes. This creates a culture where accountability is not viewed as a burden, but as a source of pride. People want to be held accountable for work that matters; the challenge is to ensure that the work they are accountable for actually does matter and that the connection is visible to everyone involved.

Building a culture of accountability is not a one-time action. It is a practice that must be reinforced continually through leadership behaviors, governance mechanisms, and communication. When accountability is embedded in the organization's DNA, it creates clarity, trust, and resilience. It ensures that data and AI initiatives are not just implemented but consistently deliver value in ways that leadership can see and the mission can feel.

Charting a Path to Lasting Impact

Every chapter in this book has reinforced a simple but powerful truth: data and AI leadership is not about technology alone. It is about connecting people, processes, and tools in ways that deliver measurable improvements to the mission. Metrics and mission integration are where all of those threads converge. They provide evidence that alignment is real, that strategy has become practice, and that technical innovation is delivering results that matter.

The work of measurement, translation, sustainment, communication, and accountability is not always glamorous. It requires discipline and persistence. Yet it is in this discipline that trust is built. Leaders who can show their organizations not only that progress has been made but also that it has created real value become indispensable. They earn the confidence of executives who control budgets, the trust of operators who rely on systems to do their jobs, and the respect of technical staff who want their work to matter.

This is not simply about capturing data points or presenting attractive dashboards. It is about cultivating a culture where evidence of value is expected, visible, and celebrated. When teams see that their work is judged not only on technical sophistication but also on mission impact, they approach problems differently. They search for relevance, they anticipate challenges, and they embrace accountability. Over time, this creates an organization that measures its progress not by activity but by results that improve outcomes.

Looking ahead, the CDAO or equivalent data leader is not only a steward of systems but also of meaning. The role is to help the organization see itself more clearly, to understand where investments are creating impact, and to use that understanding to shape future choices. This requires courage to ask the difficult questions about what truly matters, and humility to recognize when an approach must change. It requires vision to connect the technical to the strategic, and persistence to keep that connection visible long after the initial excitement of a pilot has faded.

The future of integrated impact is not one in which metrics are treated as afterthoughts or compliance checks. It is a future where measurement is woven into the fabric of daily operations, where mission outcomes are tracked with the same rigor as financial results, and where the value of data and AI is made real for everyone who depends on it. It is a future where leaders and teams see progress not as abstract potential but as visible, enduring contributions to mission success.

This is the opportunity before today's data leaders. By charting a path that integrates metrics with mission, by sustaining momentum beyond the pilot, and by building cultures of accountability, they can ensure that innovation is not just promised but delivered. They can make their organizations stronger, more agile, and more trusted. And in doing so, they can inspire others to see that the true measure of data and AI leadership is not found in the technology itself, but in the lasting impact it creates.

Chapter 9

INNOVATION ECOSYSTEM & ACQUISITION

Innovation does not arrive fully formed, nor does it appear simply because an organization declares it a priority. It is the result of deliberate design, consistent leadership, and a willingness to adapt existing systems in order to create new ones. The introduction of new technologies has always been a double-edged prospect for mission-focused organizations. On the one hand, there is the promise of new capabilities, improved efficiency, and a decisive advantage. On the other hand, there is the risk of disruption, misalignment with mission priorities, and wasted resources. The graveyard of failed technology initiatives is vast, populated by solutions that were technically sound but organizationally orphaned, brilliant in conception but impossible to procure, or promising in pilot but abandoned before scale.

This chapter explores how leaders in data and AI can bridge that gap by creating ecosystems of experimentation and acquisition that are intentionally structured to deliver value. At its heart is the recognition that innovation is not simply about tools, but about how those tools are identified, tested, and acquired. A brilliant solution sitting on a shelf has no impact if it cannot be integrated into operators' and analysts' daily work. The same is true for promising pilots that never scale into sustainable programs. Innovation must be mission-aligned and acquisition-ready, or it risks becoming another example of lost opportunity.

The Chief Data and AI Officer must therefore adopt a dual mindset. They must think like strategists, ensuring that innovation efforts are tethered to organizational goals. They must also think like operators, ensuring that new technologies can be tested quickly, acquired responsibly, and deployed at scale when they demonstrate value. This duality transforms a leader from being a passive consumer of innovation into an active broker of ideas, capabilities, and contracts. It requires fluency in acquisition mechanisms that most technical leaders have historically avoided, and comfort with ambiguity that most acquisition professionals find unsettling. The CDAO who masters both becomes indispensable.

The path forward is not to abandon rigor for the sake of speed, nor to cling to the safety of process at the expense of relevance. It is to build ecosystems of partners and acquisition models that allow organizations to move quickly while maintaining mission discipline. This is about enabling rapid experimentation, establishing agile procurement mechanisms, and cultivating emerging technology ecosystems that can sustain innovation beyond any single program or contract.

The CDAO as Connector and Catalyst

In many organizations, the Chief Data and AI Officer is still seen primarily as a technical leader or policy owner. While both roles are essential, they represent only a fraction of what the position must deliver to realize innovation at scale. The CDAO must serve as both a connector and a catalyst, actively shaping the flow of ideas, technologies, and resources among stakeholders who do not naturally speak the same language. This requires stepping beyond the boundaries of traditional executive functions and embracing the work of a broker who can translate between communities, surface hidden friction, and create momentum where processes have historically stalled.

A CDAO who cannot connect operators with technologists, or researchers with contracting officials, will quickly find that promising opportunities die in place. The brilliant algorithm developed by a research team never reaches the analysts who need it because no one navigated the data-sharing agreements. The startup with a transformative capability gives up after months of unanswered inquiries because no one told them which office to approach. The pilot that succeeded spectacularly remains a pilot forever because no one identified the contracting pathway to scale it. These failures are not technical; they are connective. The CDAO who develops trusted relationships across acquisition, compliance, program management, and mission operations can accelerate outcomes and remove friction that has historically delayed progress for months or even years.

To succeed in this role, the CDAO must first map the ecosystem of stakeholders who influence whether innovation succeeds or fails. Acquisition officials control contracting pathways and understand what mechanisms are available. Compliance officers safeguard oversight and can identify regulatory constraints before they become showstoppers. Program managers own budgets and can champion or kill initiatives based on their alignment with existing priorities. Mission operators carry the ultimate burden of effectiveness and can tell you whether a solution actually works in practice. In many cases, these groups have competing incentives and different measures of success. The CDAO's task is to surface those differences early, establish common ground, and then broker agreements that allow innovation to move forward without compromising mission integrity.

One practical approach is through structured convenings that bring acquisition, compliance, and mission voices together at the start of an innovation effort rather than at the end. In these settings, the CDAO does not need to be the subject-matter expert in each discipline, but they must be skilled in translation. When a mission operator explains a pressing capability gap, the CDAO can help acquisition officials understand how that translates into technical requirements and timeline constraints. When a technology partner describes a novel capability, the CDAO can clarify how it might align with policy constraints and governance requirements. This translation work ensures that opportunities are not lost simply because groups failed to understand one another. It also builds relationships that make future collaboration easier, because trust has been established before the pressure of a specific procurement begins.

The CDAO also acts as a catalyst, creating momentum when processes stall. Federal innovation efforts in particular often become entangled in endless cycles of pilot projects that fail to transition into operational use.

A skilled CDAO establishes clear transition criteria before experimentation begins, ensuring that all stakeholders agree on how success will be measured and on the pathways for scale. This simple discipline reduces wasted effort and increases confidence that pilots can produce more than just short-lived demonstrations. It also changes the psychology of experimentation. When teams know that success will be rewarded with resources and a path to scale, they approach pilots differently than when they suspect their work will be forgotten once the initial funding expires.

Another critical responsibility is to broker trust with external partners. Startups, academic researchers, and non-traditional vendors often find government environments intimidating and opaque. The procurement rules seem designed to exclude them. The timelines seem incompatible with their funding cycles. The feedback after unsuccessful proposals seems nonexistent. Without an advocate inside the system, these partners may never engage, or they may disengage after a single failed attempt, taking their innovations elsewhere. The CDAO can change this dynamic by establishing predictable on-ramps: standing agreements for proof-of-concept work, pre-negotiated data use arrangements, and clear explanations of what happens after an initial demonstration succeeds. By lowering the barrier to entry while maintaining compliance, the CDAO ensures that the organization has access to a richer ecosystem of innovation partners rather than being limited to the usual suspects who have learned to navigate the bureaucracy.

Ultimately, the CDAO must align innovation efforts with the broader data strategy and governance framework. Innovation pursued without connection to governance quickly creates shadow systems and compliance risks. A promising tool adopted without a proper security review becomes a liability. A dataset acquired without proper stewardship becomes a burden rather than an asset. By positioning governance as an enabler rather than a blocker, the CDAO ensures that experimentation contributes to the organization's long-term strength rather than undermining it. The connector and catalyst role is therefore not separate from the governance role; it is governance made operational, applied at the speed of relevance.

The Architecture of Experimentation

The CDAO's strength as a connector and catalyst is ultimately measured by whether good ideas reach the field. Convening the right voices is only the beginning. True innovation depends on turning alignment into action, on building the capacity to test, learn, and scale faster than the challenges evolve. This requires more than enthusiasm; it requires an architecture of experimentation that makes learning systematic rather than episodic.

Experimentation is where vision meets reality. It tests whether a promising technology or analytic method can survive contact with the mission. Too often, organizations rely on vendor claims or theoretical models, only to watch them crumble when faced with the friction of real-world operations. The data is messier than expected. The users interact with the tool in ways no one anticipated. The edge cases that seemed rare turn out to be common. Structured experimentation transforms this uncertainty into learning. It treats failure not as loss, but as data. A well-designed experiment reveals what works, what does not, and what to try next. The CDAO's role is to make such learning possible, protecting the space to test while keeping it aligned with mission relevance, safety, and accountability.

Building that capacity begins with creating environments where innovation can unfold safely. Data sandboxes and test ranges equipped with realistic, de-identified information allow teams to measure performance, surface policy issues early, and make informed adjustments before deployment. In these spaces, mission operators and technologists can work side by side, validating solutions through evidence rather than persuasion. What emerges is not only better technology, but also stronger trust between those who build and those who rely on the results. The operator who has seen a tool perform under realistic conditions becomes an advocate rather than a skeptic. The developer who has watched operators struggle with an interface learns what actually matters.

Even with the right environment, experimentation often falters on process rather than substance. Promising ideas are delayed by contracting hurdles, uncertain authorities, and slow data approvals. A team with a working prototype waits months for permission to access the data they need to validate it. A vendor with a proven commercial solution spends a year navigating procurement before they can even begin a pilot. The

CDAO's influence here is decisive. By collaborating with acquisition and legal partners, they can establish pre-negotiated frameworks for everyday needs: standing data-use agreements that can be invoked without new negotiations, model security reviews completed in advance for common architectures, and pilot-ready contract vehicles that can be activated within days rather than months. Such mechanisms do not bypass oversight; they operationalize it at the speed of relevance.

Rapid experimentation also requires discipline. Every test should begin with a clear hypothesis, measurable outcomes, and an agreed-upon decision framework. The goal is not to run pilots endlessly, accumulating demonstrations without decisions, but to create a consistent rhythm of learning that feeds action. By publishing simple, repeatable playbooks, the CDAO can help teams frame objectives, define metrics, and capture lessons in a format that enables comparison and scaling. When experiments are run this way, they produce a body of knowledge rather than a string of disconnected anecdotes. The organization learns not just from individual successes, but from the patterns that emerge across many attempts.

Measurement keeps the system honest. Tracking the time from idea to first test, from pilot to contract, and from contract to scaled capability transforms innovation from aspiration into performance. These metrics also make improvement visible. When leaders see that cycle times are shrinking, that more pilots transition to operations, and that fewer compliance issues arise late in the process, they begin to trust the system. That trust accelerates the next round of innovation, creating a virtuous cycle where success breeds confidence and confidence enables risk-taking.

Acquisition as Strategic Capability

Procurement determines whether the lessons of experimentation ever leave the lab. The most innovative technology achieves nothing if it remains trapped in contracting cycles that outlast the solution's relevance. By the time the contract is signed, the problem may have changed, the technology may have advanced, or the champions may have moved on. At the same time, moving too fast without structure can erode trust and invite risk. The answer lies in agility with accountability: procurement that is both lawful and responsive. The CDAO must therefore treat acquisition as a strategic capability, not a bureaucratic constraint to be tolerated or circumvented.

This requires fluency in the mechanisms that exist to accelerate acquisition. In the federal space, pathways such as Other Transaction Agreements, Commercial Solutions Openings, and Small Business Innovation Research programs all exist to speed the adoption of emerging technologies. Each has its own requirements and limitations, but they share a common purpose: reducing the distance between a promising idea and its operational application. Other Transaction Agreements allow partnerships with non-traditional vendors without the full weight of Federal Acquisition Regulation requirements. Commercial Solutions Openings enable organizations to solicit innovative proposals without prescribing exact solutions. SBIR programs create pipelines from early-stage research to scaled deployment. The CDAO who understands these mechanisms can match the right pathway to the right opportunity, rather than forcing every acquisition through the same slow channel.

In commercial settings, similar principles apply. Lightweight master agreements allow organizations to engage vendors quickly for initial work without lengthy negotiations. Modular contracts let teams start small, learn quickly, and scale only what proves its worth. Outcome-based contracting shifts the focus from deliverables to results, aligning vendor incentives with organizational goals. The common thread is flexibility: the ability to adapt the acquisition approach to the nature of the innovation rather than forcing the innovation to fit a rigid process.

To make innovation repeatable rather than episodic, the organization must build reusable structures. Templates for problem statements ensure that new efforts begin with a clear articulation of the need. Standardized evaluation criteria allow consistent comparison across proposals. Predefined performance metrics create shared expectations for what success looks like. When contracting language aligns with the experimental design, pilots can transition smoothly from testing to operational adoption, with predefined criteria that trigger scale-up or modification. Vendors know what success looks like. Program managers know what happens next. Compliance officers can trace the rationale for every decision. This predictability is itself a form of competitive advantage, attracting partners who want to work with organizations that respect their time and investment.

A crucial dimension of acquisition agility is openness to new partners. Startups, researchers, and small firms often hold the most creative solutions but are discouraged by complex procurement systems that seem designed for large incumbents. The CDAO can change this dynamic by offering transparency and predictability: clear engagement guides that explain how to participate, fair feedback that helps unsuccessful bidders improve, and opportunities to compete without prohibitive overhead. When the barrier to entry is lowered responsibly, the quality and diversity of ideas rise dramatically. The organization gains access to innovation that would otherwise go to competitors or remain unexploited entirely.

TREND SNAPSHOT
Federal Innovation Pathways in Procurement

Federal agencies have recognized that traditional procurement timelines are out of step with the pace of technological change. In response, they are deploying a range of agile mechanisms to accelerate acquisition while maintaining oversight.

- **Other Transaction Agreements (OTAs):** The Department of Defense used more than **$16 billion in OTAs in FY2023**, up from less than $2 billion a decade earlier. These agreements allow faster partnerships with non-traditional vendors and are frequently used for prototyping emerging technologies (GAO, 2024[74]).
- **Commercial Solutions Openings (CSOs):** Agencies such as the Defense Innovation Unit (DIU) and the Department of Homeland Security have increasingly turned to CSOs, which allow organizations to solicit and evaluate innovative proposals without prescribing exact solutions. DIU reports that more than **60 percent of its contracts go to first-time federal vendors**, broadening the pool of available technologies (DIU, 2023[75]).

74. Defense Acquisitions: Use of Other Transaction Agreements Has Increased but Oversight and Tracking Need Improvement. U.S. Government Accountability Office, GAO-24-105432, 2024.
75. Defense Innovation Unit (DIU). Annual Report 2023. U.S. Department of Defense, 2023.

- **Small Business Innovation Research (SBIR) Programs:** The Air Force's AFWERX program has restructured SBIR awards into a streamlined three-phase process, cutting contracting times from months to weeks. Since 2019, AFWERX has issued more than **4,000 SBIR contracts** to small businesses, creating a pipeline from early-stage research to scaled solutions (AFWERX, 2024[76]).
- **Challenge-Based Acquisition:** Agencies are experimenting with prize competitions and challenge models to attract non-traditional problem solvers. NASA's long-standing use of challenge acquisitions has produced breakthroughs in robotics and software (NASA 2023[77]), while DHS has used similar methods to accelerate the development of security technologies (DHS 2023[78]).

Together, these mechanisms show that federal acquisition is not static. It is evolving toward more flexible and innovation-friendly approaches, signaling to industry partners that the government is serious about reducing red tape and rewarding speed. The task for CDAOs is to harness these pathways, align them with mission outcomes, and ensure that agile procurement becomes the rule rather than the exception.

Cultivating the Innovation Ecosystem

When rapid experimentation and agile procurement become routine, they lay the groundwork for something larger: an ecosystem that sustains innovation beyond any single project or contract. No organization, no matter how well resourced, can invent or deliver every capability on its own. The pace of technological change, particularly in artificial intelligence, data management, and digital infrastructure, moves too fast for isolated effort. What endures is the network: a connected system of academic researchers, commercial partners, startups, international collaborators, and mission operators who together create an environment where ideas can be surfaced,

76. AFWERX. AFWERX Annual Report 2024. U.S. Air Force, 2024.
77. NASA. Centennial Challenges Program: Annual Report. National Aeronautics and Space Administration, 2023.
78. Department of Homeland Security (DHS). Prize Competitions and Challenges: Annual Report to Congress. U.S. Department of Homeland Security, 2023.

tested, and scaled with trust and speed.

These ecosystems are not only technical but social, built on the same principles of governance, transparency, and accountability that make experimentation possible. There is a kernel of truth in the witticism that "more business is done in one evening at the bar than in a week of a conference." Every ecosystem contains both code and culture: researchers who push the boundaries of knowledge, engineers who turn concepts into tools, mission operators who define what matters, and acquisition leaders who make it real. The CDAO's task is to hold these threads together so that innovation remains purposeful rather than diffuse. When that balance is maintained, ecosystems do more than deliver new technology; they continually renew the organization's capacity to learn, adapt, and act with confidence.

The role of the CDAO in building these ecosystems is both outward and inward. Outwardly, they extend the organization's reach, engaging universities, consortia, industry partners, and emerging ventures to surface ideas that might otherwise remain unseen. This means attending conferences not just as observers but as active participants who can articulate needs and offer partnership. It means creating structured opportunities for engagement: open challenges that invite creative solutions, problem-focused events that bring diverse perspectives to specific missions, and joint exploration sessions that allow potential partners to understand the organization's context before proposing solutions. The CDAO who waits for innovation to arrive will always be behind; the CDAO who actively cultivates external relationships will have first access to emerging capabilities.

Inwardly, the CDAO cultivates the environment that allows external ideas to take root. This means ensuring that mission operators, analysts, and technologists can engage directly and productively with external collaborators rather than being filtered through layers of bureaucracy. It means creating communities of practice that prepare internal teams to absorb new technologies as they emerge. It means building feedback loops that allow operators to signal what they need and researchers to understand whether their solutions are working. Without this inward readiness, even the best external partnerships will falter. The organization will acquire promising technologies but fail to integrate them. It will sign contracts with innovative vendors but struggle to manage the relationships. The external and internal ecosystems must develop together.

To operationalize this balance, the CDAO must begin with clarity, mapping the ecosystem that already surrounds the organization. This means identifying which vendors are under contract, which academic groups are producing relevant research, which small businesses are entering the field, and which allies or partners are developing complementary programs. Seen together, this map reveals more than a list of participants; it exposes the patterns of collaboration, the gaps where new relationships are needed, and the overlaps where effort can be shared. It also helps the CDAO recognize where the organization holds influence and where it must learn from others. With this understanding, partnerships can be shaped intentionally rather than reactively, and innovation becomes a collective pursuit instead of a series of isolated transactions.

Strong ecosystems rarely emerge on their own; they require deliberate orchestration to turn diverse contributions into coordinated outcomes. Academic researchers may bring deep expertise but limited operational reach. Startups often deliver agility but lack the stability to scale. Larger vendors possess resources and integration capacity, yet may struggle to stay close to the mission. The CDAO's task is to align each contributor to its comparative advantage and ensure that no effort operates in isolation. When orchestration works, innovation feels less like competition for attention and more like collaboration toward shared purpose. The ecosystem becomes self-reinforcing: expertise compounds, progress accelerates, and the organization learns faster than it could on its own.

Governance becomes the decisive factor that allows collaboration to flourish without descending into chaos. In a dynamic ecosystem, clear rules are not constraints but enablers. They define how partners share data, how results are evaluated, and how responsibilities are distributed. The CDAO must ensure that these frameworks are transparent, consistent, and aligned with mission priorities so that participants can focus on solving problems rather than navigating uncertainty. Governance may take the form of standardized data-sharing agreements, joint evaluation protocols, or shared playbooks for innovation management. It can also encompass intellectual property protections, security requirements, and compliance expectations that preserve trust across organizational boundaries. When governance is done well, it recedes into the background, quietly supporting collaboration and allowing creativity to operate safely at speed.

Trust is the currency that makes every ecosystem function. Startups and small innovators need to know that engagement will be fair and that their ideas will be respected. Academic partners want confidence that their research will be credited and used responsibly. Mission operators, in turn, must trust that external collaborators understand the stakes and constraints of their environment. The CDAO sits at the center of these relationships, building confidence through transparency, responsiveness, and follow-through. Even small gestures matter: providing feedback after competitions rather than silence, openly sharing evaluation criteria rather than keeping them opaque, acknowledging contributors publicly rather than treating them as anonymous vendors. Over time, these actions compound into reputation. An organization known for fairness and clarity becomes a magnet for innovation, attracting partners who want to work where ideas are valued and results are real.

VOICES IN THE FIELD
Perspectives on Building Innovation Ecosystems

"When outside partners come in, what matters most to us is that they understand the mission context. We do not need a sales pitch; we need a partner who will sit with us, listen to the problem, and adapt their technology to fit the way we operate. When that happens, we are far more willing to experiment and give honest feedback."
–Operator, U.S. Air Force

"Speed is important, but predictability builds trust. Small businesses are often wary of engaging because they are not sure what will happen after the initial conversation. When we can point to clear pathways from pilot to contract, and we keep them informed at every step, they are more likely to stay in the process and bring their best ideas forward."
–Acquisition Official, Department of Defense

"Universities have a deep bench of talent, but faculty and students need to see that their contributions will have an impact. When agencies create structured opportunities for collaboration and give recognition for research that informs mission outcomes, it makes academic labs far more willing to work through the additional requirements that come with government projects."
–Researcher, Academic Partner Institution

"Our job is to make sure the ecosystem is not fragmented. We cannot afford a situation where startups, large vendors, and academic teams are all working in parallel but never connecting. By convening them under a shared framework and keeping governance clear, we turn separate efforts into a collective force for innovation."
–CDAO, Civilian Federal Agency

From Pilots to Enduring Impact

Innovation cannot remain a series of one-time events. The goal is not to celebrate short-lived pilots or to constantly chase the next new thing, but to create pathways that turn experiments into enduring capabilities. This requires deliberate structures, committed leadership, and a culture that sees innovation not as an exception but as part of the natural rhythm of operations.

Building sustainable pathways begins with intentional design. Before an experiment starts, leaders must already have clarity on what success looks like and how it will be scaled. This does not mean knowing every detail of the eventual program, but it does mean setting clear conditions for decision-making. If an experiment meets the defined performance measures, then leaders should already have an identified contracting pathway, a budget source, and a governance framework in place to support adoption. If those measures are not met, the organization should be prepared to document lessons learned and move resources to the next opportunity without unnecessary delay. The worst outcome is ambiguity: pilots that neither succeed clearly enough to scale nor fail clearly enough to end, consuming resources and attention indefinitely.

Continuity of leadership attention is another element of sustainability. Too often, pilots begin with strong senior support but lose momentum when leaders shift to other priorities or rotate to new positions. The CDAO must ensure that innovation efforts are aligned with mission outcomes that have lasting significance, not just with the interests of current leadership. When experimentation is tied directly to operational requirements or strategic priorities that will persist regardless of personnel changes, it is more likely to survive leadership transitions and generate the institutional buy-in that sustains funding and adoption.

Workforce readiness is equally important. Sustainable innovation depends on people who know how to use new tools and are confident in their application. The most advanced technologies will falter if operators feel unprepared or unsupported. By embedding training, documentation, and communities of practice into the adoption process, the CDAO ensures that new capabilities are not just delivered but are embraced by those who need them most. This investment in human capacity is as important as the investment in technology itself. A tool that no one knows how to use is not an asset; it is a source of frustration.

The path forward is to treat innovation as an ongoing cycle rather than a series of disconnected episodes. Establish transition frameworks that connect experimentation with procurement and governance. Build feedback loops that capture lessons from both successful and unsuccessful trials. Engage mission stakeholders continuously so that new capabilities remain aligned with real-world needs. Above all, set the expectation that innovation is not a side project but a permanent feature of how the organization operates. When this expectation is established, teams approach their work differently. They think not just about whether something works, but about how it will be sustained. They build documentation and training into initial development rather than treating them as afterthoughts. They cultivate relationships with operational users rather than throwing solutions over the wall.

The promise of emerging technology is vast, but its real value is realized only when it becomes durable. The CDAO who champions sustainable pathways ensures that each experiment contributes to a larger trajectory of progress. Over time, this creates an environment where innovation is not fragile or fleeting but embedded, resilient, and generative. The organization becomes one that learns quickly, adapts continuously, and delivers results with a speed and precision that competitors struggle to match.

This is the horizon we should aim for. Not isolated pilots, not one-time demonstrations, but an enduring culture where ecosystems, procurement, and experimentation come together as a unified engine of impact. When the CDAO embraces this role, innovation ceases to be an abstract ambition and becomes a lived reality that strengthens the mission every day. The organization moves from being a consumer of innovation, waiting for others to develop solutions and then struggling to adopt them, to being a shaper of innovation, actively cultivating the partnerships and capabilities that will define its future.

Chapter 10

DEPLOYMENT PLAYBOOKS & TOOLS

The journey across this book has been about more than elevating data and AI as concepts. Each preceding chapter has explored the challenges of governance, the demands of modernization, the pursuit of ethical and responsible practice, and the necessity of aligning strategy to mission. Together, these ideas establish the scaffolding for an enterprise that understands why data matters and how leadership must respond. What remains, however, is the question of how an organization makes these ideas real in action. It is one thing to build a vision and another to equip the enterprise with the means to execute that vision consistently, reliably, and at scale.

This is where deployment playbooks and tools become central to the CDAO role. They represent the field-ready translation of concepts into methods that teams can apply under pressure and in varied circumstances. Where earlier chapters provided the intellectual foundation, the frameworks introduced here serve as the operational gear that allows people to move from knowing to doing. Playbooks and tools capture lessons learned, establish repeatable patterns, and provide the structure that prevents each new initiative from starting at zero. They are the difference between an organization that must rediscover best practices with every rotation and one that compounds its learning over time.

The need for such frameworks has surfaced repeatedly throughout this book. When we discussed modernization, we saw how outdated models of technology adoption often failed because organizations lacked shared pathways for implementation. When we examined governance, it became clear that paper policies alone do not create trust or accountability unless they are operationalized into concrete practices. When we explored innovation ecosystems, we saw that creativity without structure often leads to fragmentation and waste. In each of these contexts, the missing link was a common execution approach that could be taught, adapted, and scaled. Deployment playbooks are designed to fill that gap.

The act of equipping an enterprise with repeatable, field-ready frameworks ensures that maturity can be measured, progress tracked, and success replicated across diverse teams. More importantly, it builds organizational resilience by ensuring that lessons do not vanish when people rotate out or when leadership priorities shift. A well-built playbook carries forward institutional knowledge and keeps it accessible when needed. This is not bureaucracy; it is organizational memory made actionable.

The Nature of Playbooks

Playbooks sit at the intersection of strategy and practice. They are the mechanism by which the CDAO transforms abstract guidance into something that can be carried into the field, tested in real environments, and adjusted as lessons accumulate. Without a playbook, organizations often rely on ad hoc processes, institutional memory, or the persistence of individuals. Knowledge lives in people's heads, and when those people leave, the knowledge leaves with them. With a playbook in place, the same organization can apply a structured path that reduces variability, improves accountability, and accelerates measurable progress.

A playbook is not intended as a policy document that gathers dust on a shelf. Nor is it a rigid procedure that constrains judgment. Instead, it is a practical guide that establishes direction while leaving room for adaptation. Think of it as the difference between a map and a travel itinerary. A policy might state the destination, but a playbook provides the routes, checkpoints, and decision points along the way, giving the practitioner confidence that they are on a tested path while still allowing flexibility to respond to conditions encountered. The best playbooks feel like advice from a trusted colleague who has done this work before and wants to help you avoid the mistakes they made.

Operationally, playbooks serve several purposes. They capture hard-won lessons so that organizations do not have to repeat the same painful mistakes. A well-crafted playbook codifies proven-effective practices and flags pitfalls to avoid, sparing teams the cost of rediscovering what earlier teams already learned. They establish a shared language across teams, so that terms like stewardship, accountability, and data ownership mean the same thing to everyone. By embedding those definitions in a playbook, leaders create alignment that reduces confusion and conflict. And they serve as accelerators, providing templates, decision frameworks, and process flows that teams can adopt without reinventing the wheel. When every team is following a consistent framework, their results can be aggregated, compared, and scaled in ways that would be impossible if each team worked independently.

From a practitioner's perspective, playbooks are a safety net. They enable new leaders and staff to quickly orient themselves, even in high-tempo environments. They empower mid-level managers who may not have deep backgrounds in data or AI but are tasked with delivering results. A playbook written for responsible AI deployment, for example, can help a program manager understand the key steps to ensure fairness and transparency without requiring them to become an expert in ethics or algorithmic bias. The guidance is immediately actionable, preventing paralysis and building confidence.

Importantly, playbooks also create continuity across leadership transitions. In many government and defense organizations, rotations are inevitable. A commander serves for two years and then moves on. A program manager is reassigned just as a pilot begins to show results. Without a durable framework, progress can be lost as new leaders arrive and must restart the learning process. A playbook acts as an institutional memory, ensuring that effective practices endure even as personnel change. The new leader does not have to wonder what approach worked before; they can read it, adapt it, and build on it.

For the CDAO, the challenge is not simply to write playbooks but to craft them so they are usable in the field. That requires clarity, brevity, and direct connection to mission outcomes. A hundred-page manual filled with jargon is unlikely to be adopted, no matter how accurate it may be. A concise playbook with checklists, decision points, and examples, however, can be picked up and applied under pressure. The test of a good playbook is whether someone can use it at two in the morning when things are going wrong, and there is no one to call for help.

Assessment as Foundation

Every transformation begins with an honest look at the current state. For a CDAO, the ability to evaluate organizational maturity is not a ceremonial step; it is the foundation on which all future actions rest. Without a structured assessment, leaders often rely on assumptions, anecdotes, or incomplete narratives about how data and AI are being utilized. These narratives are frequently shaped by the loudest voices rather than by measurable evidence. The team that presents well in meetings gets resources; the team that struggles quietly gets ignored. Tools for assessment and maturity evaluation help to correct this by providing a disciplined process that surfaces reality, no matter how uncomfortable that truth may be.

Assessment tools operationalize the principles of accountability and transparency that run throughout this book. When maturity is evaluated against precise dimensions such as data quality, accessibility, analytic capability, and AI readiness, stakeholders can discuss progress without falling into debates about interpretation. The focus shifts from opinion to evidence. A good assessment tool provides both clarity and repeatability. It should be straightforward to use and repeatable over time, allowing trends to be tracked and improvements to be measured. The first assessment establishes the baseline; subsequent assessments reveal whether interventions are working.

Maturity evaluations also serve a second purpose: they connect technical conditions to mission outcomes. A low data discovery maturity score is not an abstract problem. It means that operators will waste valuable time searching for information during a mission, time that could be spent on the mission itself. A low score on AI governance does not simply reflect a lack of process; it raises the risk that an algorithm will be used in ways that undermine trust, potentially derailing adoption across the enterprise.

When assessments are framed in this way, leaders at every level can see the direct impact of maturity on the mission they are charged with advancing.

From a practical standpoint, maturity tools need to be lightweight enough to avoid becoming bureaucratic exercises, but structured enough to provide meaningful insight. Earlier attempts at large-scale data maturity models often failed because they required months of workshops and produced results that were already outdated by the time they were delivered. The organization would spend six months assessing, then another six months debating the results, and by the time action began, the world had moved on. The CDAO should resist that tendency. A well-designed tool can be administered in weeks or even days and still produce actionable findings. The key is to focus on the dimensions that matter most to the mission and to resist the temptation to measure everything simply because it can be measured.

Another crucial principle is that assessments must be repeatable and cyclical. One-off evaluations quickly become stale. By establishing a rhythm of assessment, action, and reassessment, the CDAO creates a culture of continuous improvement. This rhythm enables leaders to assess the effectiveness of interventions, compare progress across units, and identify where resources should be redirected. It also creates accountability: if the assessment showed a problem six months ago and nothing has changed, that raises questions that demand answers.

Ultimately, maturity tools should not be viewed as instruments of judgment, but rather as enablers of growth. If assessments are perceived as punitive, participants will be reluctant to provide honest responses. They will hide problems rather than surface them. If assessments are framed as tools for improvement, staff are more likely to engage openly, and leadership can leverage the results to guide investment and support. The goal is not to catch people doing things wrong; the goal is to understand reality clearly enough to make it better.

PLAYBOOK 1: Data and AI Maturity Assessment Runbook

DEPLOYABLE RUNBOOK—Customize for your organization

Use this framework to conduct a rapid assessment of organizational readiness. This runbook can be administered in workshops, surveys, or leadership interviews. Adapt questions to your specific mission context.

PRE-ASSESSMENT PREPARATION

- Identify assessment scope: entire organization, single directorate, or specific program
- Select participants: mix of leadership, technical staff, and mission operators
- Choose method: in-person workshop (2-4 hours), survey distribution (1 week), or interview series (1-2 weeks)
- Prepare scoring template: Low / Medium / High for each dimension, with space for evidence and notes

DIMENSION 1: DATA FOUNDATIONS

Evaluates whether data assets are reliable, discoverable, and appropriately governed. Weaknesses here undermine every other capability.

- Do staff have a clear and consistent way to discover data assets relevant to their mission?
- How often are data quality issues (missing values, outdated information, duplication) encountered, and how are they resolved?
- Are roles and responsibilities for data stewardship clearly defined, and are stewards actively engaged?
- What proportion of key datasets have metadata that is accurate, current, and accessible to intended users?

DIMENSION 2: INFRASTRUCTURE AND ACCESS

Examines whether the organization has the technical backbone to store, move, and secure data effectively, and whether access policies align with mission tempo.

- Are users able to access required data within mission-relevant timelines, or do approvals and technical bottlenecks create delays?
- Does the organization have a modern, scalable infrastructure that supports both structured and unstructured data?
- How are security and access controls managed, and do they balance protection with usability?
- Can the infrastructure support near-real-time data integration where the mission requires it?

DIMENSION 3: ANALYTICS AND INSIGHT GENERATION

Assesses whether the organization can transform raw data into insight that informs decision making, including both technical capability and integration into mission processes.

- Are analytic tools accessible to both technical experts and non-technical decision makers?
- How often are analytic products directly referenced in operational or leadership decision forums?
- Does the organization have repeatable processes for validating analytic methods and results?
- To what extent are advanced techniques (predictive analytics, machine learning) being piloted or scaled, and do they align with actual mission priorities?

DIMENSION 4: AI AND AUTOMATION READINESS

Assesses the organization's ability to responsibly experiment with and deploy AI. This is not a measure of how many AI systems exist, but of whether conditions are in place for sustainable and trustworthy use.

- Is there a clear framework for evaluating AI pilots, including ethical considerations and mission relevance?
- Are there policies in place that govern training data, model transparency, and accountability for AI-enabled decisions?
- Do staff have training or guidance to understand both the potential and the limitations of AI tools?
- Has the organization conducted at least one controlled pilot that demonstrates AI utility in a real mission scenario?

DIMENSION 5: ORGANIZATIONAL ALIGNMENT AND CULTURE

Recognizes that maturity is not only technical but also cultural. Focuses on leadership, workforce, and institutional alignment.

- Do senior leaders consistently communicate the importance of data and AI to mission success?
- Are incentives aligned to encourage data sharing and collaboration, or do structures reinforce silos?
- How does the organization onboard new staff into its data and AI practices, and is this process standardized?
- Have teams begun to demonstrate a culture of learning where failures in experimentation are viewed as sources of insight rather than career risk?

SCORING AND INTERPRETATION

- Score each dimension: Low (significant gaps, limited capability), Medium (foundational capability with room for improvement), High (mature capability aligned with mission needs)
- Document specific evidence for each score to support action planning
- Identify the 1-2 dimensions with greatest mission impact for priority action
- Note: Low Data Foundations + High AI Readiness = unstable foundation; address data first
- Note: High Infrastructure + Low Analytics = capability exists but is underutilized; focus on workforce and adoption

POST-ASSESSMENT ACTIONS

- Brief leadership within 2 weeks of assessment completion
- Translate priority gaps into action items using the Action Planning Template (Playbook 2)
- Schedule reassessment for 6 months to measure progress
- Archive assessment results for leadership transition continuity

Cross–reference: This framework draws on the capability evaluation principles outlined in the AI & Analytics Strategy chapter.

From Assessment to Action

An assessment on its own has limited value. It can identify weaknesses, highlight strengths, and provide a snapshot of the current state, but unless it is translated into a concrete plan of action, it risks becoming another report that sits on a shelf. Organizations are full of assessments that identify problems accurately but change nothing. The true power of assessment is realized only when it informs deliberate choices about what to do next, who will take responsibility, and how progress will be measured. Moving from assessment to action planning is therefore one of the most critical responsibilities of the CDAO.

The transition begins with prioritization. Not every gap uncovered by an assessment can or should be addressed immediately. Attempting to fix everything at once often overwhelms teams and dissipates resources. The organization ends up doing many things poorly rather than a few things well. Instead, the CDAO should work with stakeholders to identify the most mission-critical areas where action will generate a visible impact. These may be the issues that most directly affect readiness, decision speed, or trust in data. If assessments reveal that data discovery is poor but infrastructure is generally stable, the first priority might be to implement a catalog or search capability to reduce wasted time finding information. This focus creates a clear win that directly supports operational outcomes and builds credibility for subsequent initiatives.

Once priorities are identified, the next step is to define actions with specificity. General statements such as "improve data governance" or "expand AI use" are not actionable. They provide direction without traction. An effective action plan breaks these down into concrete steps, such as "establish a cross-functional data stewardship group for logistics data" or "pilot an AI-enabled forecasting model in one regional command." Each action should have a defined owner, a timeline, and a clear metric of success. This level of detail transforms abstract goals into commitments that can be executed and monitored. It also enables accountability: if the action has an owner and a deadline, someone can be asked whether it happened.

Action plans should be designed with flexibility in mind. The world does not hold still while plans are executed. Each initiative should be structured so that it can be scaled up if successful, modified if conditions change, or shut down if it proves ineffective. This prevents the organization from becoming trapped in rigid plans that no longer match reality. The pilot that seemed promising six months ago may have revealed fundamental problems; the action plan should allow for that learning to change the course. Flexibility is not the same as vagueness. The plan should be specific about what success looks like while remaining adaptable about how to get there.

Communication is another essential element. Assessments can create anxiety, especially if they highlight shortcomings. Teams that have been working hard may feel that an assessment is criticizing their efforts. By presenting the action plan as a constructive path forward, the CDAO shifts the narrative from critique to progress. Leaders should communicate not only the identified problems but also the actions being taken and the expected benefits for the mission. This helps maintain trust and signals to the workforce that assessments are not punitive but part of an improvement cycle.

The final step is to establish a rhythm of reassessment. Action plans are not static. They must be revisited regularly to determine whether they are producing the intended results. This cycle of assessment, action, and reassessment creates a feedback loop that drives continuous improvement. It also provides evidence of progress that can be communicated to leadership and external stakeholders. Over time, this rhythm builds confidence in the CDAO function by demonstrating that the office not only diagnoses problems but also drives measurable results.

PLAYBOOK 2: Action Planning Template

DEPLOYABLE RUNBOOK—Customize for your organization

Use this template to translate assessment findings into executable initiatives. Complete one template for each priority action. Maintain a portfolio of active action plans for leadership visibility and transition continuity.

1. ACTION TITLE

A concise, descriptive statement of the initiative.

Example: "Implement a searchable data catalog for logistics datasets."

[Your action title]: ______________________________

2. RATIONALE AND LINK TO ASSESSMENT

Which assessment finding does this action address?

Why was it prioritized over other gaps?

Example: "The maturity assessment identified data discovery as a critical weakness. Operational units reported losing hours locating relevant datasets. This action directly addresses that gap and links to readiness outcomes."

[Your rationale]: ______________________________

3. ACTION OWNER

Identify the single individual or office accountable for delivering this action. Shared ownership is no ownership.

Example: "Chief of Logistics Data Division."

[Your owner]: ______________________________

4. SUPPORTING ROLES

List other stakeholders who will contribute or provide input. This builds clarity on who is involved without diluting accountability.

Example: "IT Infrastructure Branch for hosting, Governance Council for metadata standards, Operations Directorate for pilot feedback."

[Your supporting roles]: ______________________________

5. KEY STEPS

Outline the sequence of activities required. Be specific enough to guide execution but flexible enough to allow adaptation.

Example:

- Select a platform for cataloging logistics data
- Identify and onboard initial priority datasets
- Define metadata standards in coordination with governance
- Conduct a 60-day pilot with one operational unit
- Refine and scale catalog based on pilot feedback

[Your key steps]:

Step 1: ______________________________

Step 2: ______________________________

Step 3: ______________________________

Step 4: ______________________________

Step 5: ______________________________

6. TIMELINE

Provide a realistic time frame with milestones. Ambitious enough to create momentum but not so aggressive that it is unachievable.

Example: "Initiation in Month 1, pilot completed by Month 3, full operational rollout by Month 6."

[Your timeline]: ______________________________

7. METRICS OF SUCCESS

Define measurable indicators tied to mission value, not just technical completion.
Example:

- At least 80% of logistics staff in pilot unit able to locate required datasets within 15 minutes
- Reduction in redundant requests for data by 50% within 90 days of catalog deployment

[Your metrics]:

Metric 1: ____________________

Metric 2: ____________________

8. RISKS AND MITIGATIONS

List foreseeable risks and how they will be managed.
Example:

- Risk: Users resist adoption. Mitigation: Provide short training sessions and embed catalog use into reporting requirements.
- Risk: Metadata quality inconsistent. Mitigation: Assign stewardship responsibilities for each dataset with review checkpoints.

[Your risks and mitigations]:

Risk 1: ____________________

Mitigation: ____________________

Risk 2: ____________________

Mitigation: ____________________

9. STATUS AND REVIEW CYCLE

How often will progress be reviewed and by whom?

Example: "Monthly check-ins with CDAO office, quarterly progress report to senior leadership."

[Your review cycle]: ______________________________

PORTFOLIO MANAGEMENT NOTE

Maintain all active action plans in a single portfolio document. Review the portfolio monthly to track progress across initiatives, identify resource conflicts, and communicate status to leadership. Archive completed actions with lessons learned for future reference.

Onboarding and Institutional Memory

One of the most consistent challenges organizations face—particularly in government and defense contexts—is personnel turnover. Rotations, reassignments, and leadership changes follow a predictable cycle, and with every change comes the risk of losing institutional knowledge. The commander who finally understood the data landscape moves on; the analyst who knew where everything was stored gets reassigned; the program manager who championed the AI pilot retires. For a CDAO, this reality cannot be treated as a minor inconvenience. It must be seen as a fundamental operational challenge. Without a plan to quickly absorb new personnel and integrate them into the data and AI ecosystem, the enterprise risks losing momentum and repeating costly mistakes.

Structured onboarding is therefore not optional. It is central to sustaining progress. Onboarding is not simply about showing a new staff member where to log in or which meetings to attend. It is about immersing them in the mission context, connecting them with the right resources, and equipping them with tools that shorten the learning curve from months to weeks. The goal is to make new personnel productive as quickly as possible while reducing the risk that essential practices are lost in transition.

A CDAO-led onboarding program should cover several layers. At the most basic level, it must orient new staff to the technical environment: where data resides, how to request access, and what systems are considered authoritative. Beyond the technical layer, it should explain the governance model so that every staff member knows who is responsible for stewardship, how data is classified, and where accountability sits. Without that context, staff may unintentionally undermine governance simply by not knowing the established processes. They may create duplicate datasets because they did not know an authoritative source existed. They may share data inappropriately because no one explained the classification rules. Finally, onboarding must introduce the organization's cultural expectations. If data sharing is a priority, that value must be communicated directly and reinforced through early assignments. If experimentation with AI is encouraged, staff should be introduced to the frameworks that allow them to do so responsibly.

Rapid integration is also about equipping new leaders to act decisively. When a commander or senior program manager arrives in a new role, they do not have the luxury of spending six months learning the landscape. They are expected to make decisions from day one, and those decisions will be better informed by an accurate understanding of the data and AI environment. The CDAO can provide structured briefing materials that link the state of data and AI directly to mission outcomes. These materials should include maturity assessment results, current action plans, and the key risks or opportunities that the new leader needs to know immediately. In this way, onboarding becomes a strategic accelerant rather than a bureaucratic formality.

When combined with rapid integration tools such as maturity dashboards, role-specific playbooks, and clear access pathways, structured onboarding becomes a multiplier. It ensures that new personnel not only join the organization but also integrate into its mission tempo at speed. This capability strengthens resilience, builds trust, and helps sustain the momentum of modernization and innovation, even when the people carrying the work forward inevitably change. The playbook outlives the person who wrote it; that is its purpose.

PLAYBOOK 3: First 30 Days CDAO Onboarding Guide

DEPLOYABLE RUNBOOK–Customize for your organization

Use this playbook to structure the first month for a new CDAO or senior leader entering an organization where data and AI are critical to the mission. Prepare materials in advance so incoming leaders can execute this sequence immediately upon arrival.

WEEK 1: ORIENTATION AND CONTEXT

Priority: Gain clarity on the environment. Filter the flood of information and focus on what truly matters.

Mission Alignment:

- Review the organization's mission documents and strategic priorities
- Note how data and AI are referenced (or absent) in existing strategies
- Identify the top 3 mission priorities that data/AI should support

Governance Frameworks:

- Map the current governance model: Who owns data stewardship responsibilities?
- Identify where policy authorities reside
- List active working groups or councils related to data/AI

Access and Infrastructure:

- Confirm access to systems, dashboards, and analytic environments
- Identify bottlenecks that might slow early engagement
- Request credentials for all relevant platforms by Day 3

Key Relationships:

- Schedule meetings with: CIO, CTO, Chief of Staff, operational leaders
- Ask each: "What is your biggest data/AI pain point?"
- Document initial impressions and concerns raised

WEEK 2: BASELINE AND ASSESSMENT

Priority: Establish a clear picture of the current state using maturity assessment tools.

Maturity Assessment:

- Review most recent maturity evaluation results (use Playbook 1 if none exists)
- Examine scores across: data foundations, infrastructure, analytics, AI readiness, culture
- If no recent assessment exists, conduct rapid interviews using Playbook 1 questions

Gaps and Strengths:

- Identify 1-2 areas where the organization is clearly strong
- Identify 1-2 areas where immediate challenges are visible
- Document specific evidence for each assessment

Operational Risk Linkages:

- Ask: How are these gaps influencing mission outcomes today?
- Example questions: Are intelligence teams slowed by lack of access? Is analytic insight being ignored in decision forums?

Cultural Temperature Check:

- Hold 3-5 short listening sessions with staff across levels
- Gauge trust levels, enthusiasm, and sources of resistance
- Note: What do people complain about? What do they wish existed?

WEEK 3: ENGAGEMENT AND PRIORITIZATION

Priority: Shift from listening to engaging. Identify quick wins and establish credibility through early action.

Quick Wins:

- Identify at least one initiative demonstrable within 90 days
- Examples: standardized data request process, simple dashboard tied to mission priority, clearing a known access bottleneck
- Select something visible, achievable, and connected to a pain point heard in Week 1-2

Stakeholder Mapping:

- Build a working map of formal and informal influence over data/AI decisions
- Identify allies who share your vision
- Identify potential resistance and their concerns

Policy Familiarization:

- Review active policy directives: data management, cybersecurity, AI experimentation
- Note gaps or outdated guidance requiring update
- Flag any compliance risks requiring immediate attention

Playbook Introduction:

- Introduce deployment playbooks concept to peers and subordinates
- Signal that structured tools will be part of the operating model
- Share this onboarding guide as an example of the approach

WEEK 4: ACTION PLANNING AND COMMUNICATION

Priority: Set direction. Turn accumulated knowledge into a clear action path.

Ninety-Day Plan:

- Draft action plan with 3-5 specific goals using Playbook 2 template
- Each goal should connect to maturity gaps identified in Week 2
- Each goal should be achievable within the next quarter

Communication Strategy:

- Develop a narrative to share with the workforce
- Key message: Data and AI are enablers of the mission, not distractions
- Celebrate early strengths while being candid about growth areas

Leadership Engagement:

- Share action plan with senior leadership for endorsement
- Ensure alignment with organizational priorities
- Secure commitment for resources needed to execute quick wins

Integration of Lessons:

- Capture what you learned during the first month
- Update this onboarding playbook with organization-specific details
- Prepare materials for the next leader who will follow you

ONBOARDING PACKET CHECKLIST

Prepare these materials in advance for any incoming leader:

- Standardized briefing packet on data/AI landscape
- Latest maturity assessment results
- Summary of governance roles and authorities
- Stakeholder map (formal and informal)

- Active action plans portfolio
- Access request forms and system credentials checklist
- This 30-day onboarding playbook

Cross–reference: This playbook aligns with the capability evaluation approach in the AI & Analytics Strategy chapter and connects to Playbooks 1 and 2 in this chapter.

Enabling Rapid Mission Alignment

The central purpose of deployment playbooks and tools is not simply to create organizational efficiency or technical consistency. Their true value lies in how they allow the enterprise to align quickly with the mission. Alignment is the point at which efforts in data and AI stop being background activities and begin to shape real operational outcomes. For the CDAO, this alignment must be more than rhetorical. It must be demonstrated in visible ways that leaders, operators, and partners can see in their own work.

Rapid mission alignment requires three deliberate steps. The first is translation. Leaders must translate the language of mission priorities into data and AI objectives. If a command's top priority is improving readiness, the translation might focus on reducing delays in maintenance reporting by improving data integration. If a health agency emphasizes early detection of outbreaks, the translation could be framed as improving analytic surveillance across disparate datasets. By making this translation explicit, the CDAO ensures that every initiative has a direct line of sight to mission outcomes. The connection is not assumed; it is documented and communicated.

The second step is synchronization. Even when objectives are clear, different parts of the organization often pursue them in isolation. The logistics team builds one solution, the personnel team builds another, and the intelligence team builds a third, all addressing similar problems in incompatible ways. Playbooks help synchronize activity by providing a shared process that multiple teams can follow. A playbook for AI experimentation can be adopted simultaneously by different units, ensuring that pilots are executed consistently and that lessons can be shared across domains. Synchronization also enables scaling success. If one unit demonstrates value with a small pilot, the framework exists to replicate it elsewhere without losing fidelity or reinventing the approach.

The third step is acceleration. Once translation and synchronization are in place, the CDAO uses action-planning and assessment cycles to accelerate progress. This involves setting near-term milestones that demonstrate value quickly while keeping the long-term trajectory in mind. A capability that arrives too late, no matter how advanced, fails to serve the mission. The organization that can move from concept to pilot to scale in months rather than years gains a decisive advantage. Playbooks enable this acceleration by reducing the time spent figuring out how to proceed and increasing the time spent actually proceeding.

Communication is essential throughout this process. Alignment is not achieved silently. The CDAO must actively tell the story of how data and AI initiatives are supporting the mission. This does not mean exaggerating claims or overselling early pilots. It means showing leaders and staff where the work is making a difference, even if those differences are incremental. Sharing concrete examples of faster decision cycles, improved accuracy, or reduced costs builds trust and creates momentum. Without this narrative, alignment risks remaining invisible, which can erode support and slow adoption. The work may be real, but if no one knows about it, the political capital needed for the next initiative will not exist.

In many ways, rapid mission alignment is the culmination of the themes addressed throughout this book. Modernization gave us the technological foundation. Governance provided the rules and accountability structures. Innovation ecosystems created pathways for experimentation and creativity. Deployment playbooks bring all of these together in a way that is focused, synchronized, and directly tied to the mission. By enabling rapid alignment, the CDAO ensures that data and AI are not abstract concepts but living capabilities that advance the organization's goals in real time.

PLAYBOOK 4: Mission Alignment Quick-Start

DEPLOYABLE RUNBOOK—Customize for your organization

Use this runbook to translate organizational mission priorities into data and AI initiatives. Complete this exercise quarterly or when mission priorities shift.

STEP 1: CAPTURE MISSION PRIORITIES

List the organization's top 3-5 stated mission priorities. Source these from strategic plans, commander's guidance, or leadership communications.

Priority 1: ______________________________

Priority 2: ______________________________

Priority 3: ______________________________

Priority 4: ______________________________

Priority 5: ______________________________

STEP 2: TRANSLATE TO DATA/AI OBJECTIVES

For each mission priority, identify how data or AI could directly support it. Be specific about the mechanism.

Example: Mission Priority = "Improve readiness" Data/AI Objective = "Reduce maintenance reporting delays through automated data integration"

Priority 1 → Data/AI Objective: ______________________

Priority 2 → Data/AI Objective: ______________________

Priority 3 → Data/AI Objective: ______________________

Priority 4 → Data/AI Objective: ______________________

Priority 5 → Data/AI Objective: ______________________

STEP 3: ASSESS CURRENT INITIATIVES

List all active data/AI initiatives. For each, note which mission priority it supports (if any).

Initiative: ____________________ Supports Priority: ____________________

Initiative: ____________________ Supports Priority: ____________________

Initiative: ____________________ Supports Priority: ____________________

Initiative: ____________________ Supports Priority: ____________________

Flag any initiatives that do not clearly connect to a mission priority for review.

STEP 4: IDENTIFY GAPS

Which mission priorities lack data/AI support? These represent opportunities for new initiatives.

Gap 1: ____

Gap 2: ____

STEP 5: SYNCHRONIZE ACROSS UNITS

Identify which organizational units are working on similar objectives. Look for opportunities to share approaches, consolidate efforts, or adopt common playbooks.

Overlapping efforts: ____

Synchronization opportunity: ____

STEP 6: SET ACCELERATION MILESTONES

For priority initiatives, set near-term milestones (30, 60, 90 days) that will demonstrate visible progress.

Initiative: ____________________ 30-day milestone: ____________________

Initiative: ____________________ 60-day milestone: ____________________

Initiative: ____________________ 90-day milestone: ____________________

STEP 7: COMMUNICATE THE CONNECTION

Draft a brief narrative (2-3 sentences) explaining how each priority data/AI initiative supports mission outcomes. Use this in leadership briefings and workforce communications.

Narrative: ____

____

____

Review this alignment map quarterly. As mission priorities evolve, data/AI initiatives should evolve with them.

Agility as Enduring Advantage

The ultimate measure of a data and AI program is not the number of initiatives launched, tools procured, or playbooks written. It is found in the organization's ability to act quickly, precisely, and confidently when the mission demands it. Deployment playbooks and tools serve as the foundation of that capability. They transform data and AI from abstract ambitions into lived practices that can be carried into the field, repeated under pressure, and scaled across diverse contexts.

Throughout this book, we have seen how modernization, governance, accountability, and innovation ecosystems provide the scaffolding of transformation. Each of these elements is necessary, but none of them guarantees impact on its own. The final piece is operational agility: the capacity to bring together people, process, and technology in ways that respond to shifting priorities without losing coherence or momentum. Agility is not improvisation. It is the result of deliberate preparation, repeatable frameworks, and a culture that values learning as much as performance.

Playbooks and tools enable this kind of agility. They standardize what should be consistent while leaving room for adaptation where flexibility is required. They shorten the learning curve for new staff and leaders, ensuring that progress does not stall when roles rotate. They embed lessons from past successes and failures into the daily rhythms of work, so that knowledge becomes institutional rather than individual. Most importantly, they ensure that the energy of experimentation and innovation is always directed back toward mission outcomes. Innovation for its own sake is a luxury; innovation in service of the mission is a necessity.

For practitioners, the aspiration of agility should not feel abstract. It should be visible in everyday actions. When a new project begins, a playbook provides a tested starting point rather than forcing teams to reinvent the process. When an assessment identifies gaps, an action-planning tool helps translate them into practical steps with owners and timelines. When leaders demand evidence of progress, dashboards and maturity models provide clear and transparent reporting. Each of these examples represents agility in action, allowing the organization to move quickly without sacrificing structure or accountability.

Agility allows the enterprise to adapt to disruption, to absorb the shocks of turnover or shifting missions, and to capitalize on opportunities without being paralyzed by uncertainty. It builds not only resilience but also an enduring advantage, because an organization that can adapt faster than its challenges will always be positioned to succeed. The competitors who spend months figuring out how to start will still be planning when the agile organization has already learned, adjusted, and moved on.

If modernization is about creating the conditions for progress, and governance is about ensuring accountability, then operational agility is about making sure that progress endures. It is the difference between short-lived initiatives and lasting transformation. With playbooks and tools at the center of the approach, agility becomes more than a goal. It becomes a defining characteristic of the organization, one that ensures it is always prepared to meet the mission, no matter how it evolves.

PLAYBOOK 5: Creating Your Own Playbooks

DEPLOYABLE RUNBOOK–Customize for your organization
Use this meta-playbook to develop additional runbooks for your organization. Every recurring challenge is an opportunity for a playbook that captures lessons and accelerates future efforts.

STEP 1: IDENTIFY THE RECURRING CHALLENGE
What process or situation occurs repeatedly and would benefit from standardization?
Examples: AI pilot evaluation, data sharing agreement negotiation, dashboard deployment, vendor assessment, incident response

Challenge to address: ______________________________

____

STEP 2: GATHER LESSONS LEARNED
Interview 3-5 people who have navigated this challenge before. Ask:

- What worked well?
- What would you do differently?
- What do you wish someone had told you at the start?
- What are the common failure modes?

Key lessons: ______________________________

____

STEP 3: DEFINE THE SEQUENCE

Break the challenge into discrete steps. Each step should be:

- Specific enough to guide action
- Flexible enough to adapt to context
- Connected to a clear output or decision point

Draft sequence:

Step 1: ____

Step 2: ____

Step 3: ____

Step 4: ____

Step 5: ____

STEP 4: ADD DECISION POINTS AND CHECKPOINTS

Where should the user pause to evaluate progress or make a choice?

After Step ________________: Decision point–____________________

After Step ________________: Checkpoint–_______________________

STEP 5: INCLUDE EXAMPLES AND TEMPLATES

What examples, templates, or fill-in-the-blank sections would help users apply the playbook?

Examples to include: ________________________________

Templates to include: _______________________________

STEP 6: PILOT AND REFINE

Test the playbook with one team or project before broad distribution.

Pilot team/project: ____

Feedback collected: ____

Revisions made: ____

STEP 7: PUBLISH AND MAINTAIN

Make the playbook accessible and establish a review cycle.

Publication location: ____

Review cycle: Annually or after significant lessons learned

Owner responsible for updates: _________________________________

PLAYBOOK QUALITY CHECKLIST

- Can someone use this at 2 AM when things are going wrong?
- Is it concise enough to be read in under 15 minutes?
- Does it include specific examples, not just abstract guidance?
- Does it leave room for judgment and adaptation?
- Is ownership clear for maintaining and updating it?
- Does it connect to mission outcomes, not just process compliance?

Conclusion

THE FUTURE OF THE CDAO ROLE

The role of the Chief Data and AI Officer has been defined by change since its inception. It is not a role created to maintain the status quo, nor is it one that can be carried out through a narrow technical lens. The CDAO and other adjacent roles were born out of necessity: a recognition that data and artificial intelligence were reshaping the operating environment of every institution at a pace faster than most organizations could adapt. Throughout this book, we have explored how this role evolved, its underlying foundations, and the various ways it can drive transformation when equipped with the proper authority, structure, and resources. This final chapter asks a different question: **"What comes next?"**

The answer is not simply a continuation of what we have already seen. It is the culmination of every thread we have traced together. We began with the strategic and policy foundations that establish the CDAO's authority to act. We moved through governance frameworks that balance control with agility, modernization efforts that liberate organizations from technical debt, and organizational designs that allow new roles to take root and flourish. We examined how AI and analytics strategies translate ambition into capability, how responsible practice ensures that power is wielded with wisdom, and how innovation ecosystems bring fresh energy into enterprises that might otherwise calcify. We explored the discipline of metrics and mission integration (the crucial work of proving that data and AI actually deliver results) and equipped practitioners with deployment playbooks that translate concepts into daily action. Each of these chapters represents a foundation stone. Together, they form the platform upon which the future CDAO will stand.

The CDAO is not merely a new executive title to be added to an organizational chart. It is becoming an institution that reflects the realities of the modern age. Just as the emergence of the Chief Financial Officer marked a turning point in the professionalization of financial management, and just as the Chief Information Officer became a fixture during the rise of enterprise computing, the CDAO is poised to define the era of intelligent, data-driven organizations. The journey so far has been about building a strong foundation. The journey ahead is about institutionalizing that progress, scaling it across missions and enterprises, and ensuring it endures through cycles of leadership change, market disruption, and technological upheaval.

The Convergence of Forces

The future of the CDAO is not being shaped by a single trend or by the emergence of a single new technology. It is defined by the convergence of multiple forces unfolding simultaneously, and the velocity of their interaction makes this role so indispensable. Throughout this book, we have traced many of these forces separately: the need for alignment between strategy and policy, governance models that foster trust, modernization efforts that eliminate outdated practices, and innovation ecosystems that bring new energy into the enterprise. We have examined the cultural dimensions of leadership, the importance of ethical guardrails, and the discipline of measuring impact in ways that connect directly to mission outcomes. Taken together, these forces are not distinct lanes but currents feeding into a single river. The CDAO is positioned at the center of this flow, responsible for guiding the organization as it navigates the turbulence and opportunities that arise at this intersection.

On the technology front, we are experiencing breakthroughs at a pace that feels almost disorienting. Generative models and advanced automation tools are only the most visible examples. They are accompanied by advances in cloud architectures, edge analytics, large-scale knowledge graphs, and digital ecosystems that allow data to move securely across organizational boundaries. These are not isolated capabilities but building blocks that change expectations about what organizations can accomplish. As we discussed in the chapters on modernization and AI strategy, technology alone has never been enough. It must be paired with the right structures and mindsets. The CDAO of the future must act as the integrator who ensures that new technical possibilities translate into practical, repeatable, and responsible improvements in mission performance.

At the organizational level, expectations are also shifting. Decision makers no longer view data and AI as experimental or optional. Leaders now want these capabilities woven into daily operations, not simply piloted in a lab or highlighted in an annual report. This shift toward everyday modernization and everyday governance, a theme that recurred throughout this book, reflects a recognition that transformation is not a project with a completion date but a discipline that must be sustained indefinitely. As technology capabilities rise, so too does the demand from executives, operators, and the public for tangible outcomes that demonstrate both speed and reliability. The CDAO stands at the center of this expectation cycle, accountable for building confidence that investments in data and AI yield real value.

There is also a societal force at work. Trust, fairness, and transparency are no longer viewed as abstract ideals but as measurable requirements. The chapter on ethics and responsible practice demonstrated how these considerations can be integrated into daily workflows rather than bolted on afterward. Citizens, employees, and partners expect that the use of data and AI will be both effective and just. This expectation will only grow as these technologies become more visible in decisions that affect healthcare, justice, defense, education, and economic opportunity. The CDAO will be asked to stand as a steward of this trust, ensuring that every deployment of AI or expansion of data access is not only lawful but defensible in the court of public perception.

What makes this convergence so challenging is that these forces do not move on the same timeline. Technology evolves by the quarter.

Organizational change often unfolds over the year. Cultural and societal expectations can span decades. The CDAO must learn to operate across these time horizons simultaneously, making choices that deliver near-term wins for leadership while building structures that can endure long after the next technology cycle or leadership turnover. In this sense, the CDAO is not only an operator but also an architect of the future, carrying forward lessons from what has worked and what has failed while preparing the ground for the next chapter of organizational life.

The CDAO as Living Institution

The CDAO is not a temporary experiment or a passing title created in response to a trend. It is a role that is steadily becoming institutionalized, woven into the enduring fabric of how organizations lead, govern, and deliver outcomes. We have seen throughout this book how CDAOs are tasked not only with managing datasets or building analytical tools, but also with creating environments where innovation can take root, governance can balance control with flexibility, and ethical practices can guide deployment at scale. These are not temporary responsibilities. They are the hallmarks of an institution that is forming in real time.

The strength of this role lies in its ability to translate across multiple domains: strategy and policy, governance and operations, culture and technology. The future of the CDAO depends on institutionalizing this translation capability so that it becomes integral to every organization's DNA. Just as the CFO role emerged when financial complexity reached a point where expertise had to be formalized, and just as the CIO became a fixture during the rise of enterprise computing, the CDAO is now emerging as the institutional anchor for a world where data and AI are central to every mission.

Institutionalization, however, is not automatic. It requires deliberate action and the establishment of precedent. The governance models chosen today will set the expectations for how future CDAOs strike the balance between agility and accountability. The structures built for modernization will become the templates by which others assess what is possible and sustainable. The approaches to measuring impact, described in our chapter on metrics and mission integration, will become the benchmarks that define what success looks like for years to come. Every decision by a current CDAO carries weight beyond the immediate mission. These decisions ripple outward, shaping not just local outcomes but the broader ecosystem of practice.

There is also a cultural dimension to institutionalization. A role becomes an institution not only through policies and structures, but through norms and shared understanding. The CDAO helps build a culture where data is treated as a strategic asset, where AI is deployed responsibly, and where learning from failure is seen as progress rather than weakness. The chapter on organizational design and change leadership highlighted that cultural adoption is what transforms programs into lasting capability. The same principle applies here: the CDAO becomes a living institution when the values they champion take hold across the workforce and endure through leadership transitions.

The idea of the CDAO as a living institution also demands investment in people. No institution is sustained by one leader alone. It requires the growth of a community, the establishment of career pathways, and the deliberate cultivation of talent. This includes developing deputy CDAOs, field CDAOs, and domain-specific specialists who bring expertise in areas such as ethics, human-machine teaming, and cross-domain data integration. As we discussed in the chapter on innovation ecosystems, the health of any system depends on a diverse and connected network of contributors. The CDAO will endure as an institution only if it is supported by a pipeline of capable leaders who are given the training, authority, and trust to carry the mission forward.

Toward the AI-Ready Enterprise

What does it mean to become AI-ready? This book has framed readiness not as a final milestone but as a journey that demands alignment between policy, governance, technology, and culture. To be AI-ready is to create the conditions where data and artificial intelligence are not occasional tools but enduring capabilities, integrated into every corner of the enterprise. The future of the CDAO will be judged on how well organizations can make this vision real.

Building an AI-ready enterprise begins with the work described in this book. Strategy and policy alignment provide the legitimacy and authority required to act decisively. Governance ensures that authority is exercised responsibly and consistently. Technology modernization enables organizations to unlock legacy data, dismantle silos, and operate at the speed demanded by mission realities. Organizational design and change leadership establish the structures and incentives that allow transformations to take root. Metrics and mission integration provide the feedback loops that enable leaders to measure progress and refine approaches. Taken together, these foundations form the conditions for building an AI-ready enterprise.

To carry this forward into the future requires three interlocking outcomes that together define what it means to be truly AI-ready.

The first is **everyday integration**. Throughout this book, we have spoken of everyday modernization and everyday governance—the imperative to integrate data and AI into daily workflows rather than treating them as one-time projects or isolated experiments. An AI-ready enterprise is one in which advanced analytical tools, machine learning models, and intelligent assistants are not reserved for a select few specialists. They are accessible to operators, decision-makers, and frontline staff in a way that feels natural and trustworthy. The CDAO ensures that AI becomes a seamless extension of human work, augmenting rather than replacing, clarifying rather than complicating.

The second is **adaptive governance**. Governance that is too rigid will stifle innovation, yet governance that is too loose will erode trust. Our earlier exploration of governance frameworks showed that organizations must strike a balance that enables agility without sacrificing accountability. For the AI-ready enterprise, this balance must be maintained across multiple time horizons and risk levels. As new technologies emerge, policies and guardrails must evolve in step. As societal expectations evolve, transparency and ethical safeguards must be recalibrated accordingly. The future CDAO will not view governance as a static framework but as a living system that adapts as quickly as the environment demands.

The third is **mission agility**. The chapters on innovation and deployment emphasized the importance of being able to pivot quickly when circumstances change. In a world where new threats emerge without warning and new opportunities appear without long lead times, agility becomes the defining competitive advantage. For the AI-ready enterprise, agility does not mean chasing every trend. It means having the foundation in place to adapt rapidly when adaptation matters: reconfiguring data pipelines to address new operational requirements, deploying AI models in days rather than months, or realigning organizational priorities when mission conditions shift. The CDAO becomes the architect of this agility, ensuring that organizations are not locked into rigid structures but are prepared to move quickly and precisely when the situation demands it.

These outcomes are not theoretical. They are already emerging in organizations that have embraced the lessons discussed throughout this book. We see it in defense contexts where mission operators have access to real-time insights powered by AI. We see it in healthcare systems where predictive analytics guide patient care and resource allocation. We see it in financial and regulatory environments where AI models detect anomalies and enhance oversight without overwhelming human staff. In each case, the thread that runs through success is leadership that understands how to make AI both practical and trustworthy. That leadership is embodied in the CDAO.

Global and Cross-Sector Leadership

The CDAO is not confined to a single organization, sector, or nation. The forces shaping this role are inherently global and cross-cutting. Data does not respect traditional boundaries, and artificial intelligence is advancing in ways that connect markets, governments, and societies, whether leaders are prepared or not. The future of the CDAO will increasingly be defined by the ability to operate across sectors and borders, forging common ground where the pressures of competition and the demands of collaboration intersect.

When discussing innovation ecosystems, we noted that progress accelerates when industry, government, and academia share knowledge and resources. When exploring governance, we observed that interoperable standards are more effective, enabling different communities to collaborate with confidence. These principles now extend to the global stage. The CDAO is not simply a leader within a single enterprise. They are part of a broader movement building shared capacity across industries and nations.

This shift has practical implications. In the commercial sector, organizations are experimenting with AI to enhance efficiency, personalize services, and tap into new markets. In government, AI is being applied to defense, health, transportation, and citizen services. In academia, research is advancing the boundaries of what these technologies can do while also raising critical ethical questions. Each of these domains is learning in real time, but none can succeed in isolation. The future CDAO must act as a bridge across these communities, identifying where lessons can be transferred, where standards can be harmonized, and where collaboration is essential.

There is also an international dimension. Nations are investing heavily in data and AI capabilities, each seeking an advantage in innovation, security, and economic growth. At the same time, there is recognition that no nation can act alone. Cross-border data flows underpin global commerce. Shared scientific challenges (climate change, public health, cybersecurity) demand cooperative use of data and AI. Ethical debates surrounding transparency, bias, and accountability are ongoing in every society. The CDAO becomes a critical actor in this international dialogue, representing not only their organization but also the values and practices of their sector or nation. In doing so, they help shape the global norms that will govern how data and AI are developed and applied for decades to come.

The institutionalization of the CDAO will not stop at the borders of individual organizations. It will expand into a profession with shared practices across sectors and shared influence across nations. The CDAO will become a recognized leader in international dialogues, much as CFOs and CIOs became central figures in shaping global standards for finance and technology. This is not a theoretical possibility. It is already happening, and the pace will only accelerate.

Inspiring the Next Generation

No institution lasts without renewal, and no movement succeeds unless new leaders are inspired to step forward. The CDAO role cannot remain in the hands of a small circle of pioneers. For this role to endure as an institution, there must be a deliberate effort to prepare the next generation—those who will carry the mission farther than those who came before. This is not a secondary task. It is central to ensuring that the work of today outlasts the tenure of any one leader.

Throughout this book, we have described the CDAO as a translator—someone who bridges the worlds of policy and strategy, governance and operations, technology and mission. That act of translation must also extend across generations. Today's CDAOs have a responsibility to make their decisions visible, to share the reasoning behind their choices, and to pass forward not only their successes but also their setbacks. In this sense, the CDAO becomes a part-time historian. Each initiative, governance model, and modernization effort becomes a record that future CDAOs can study to understand context and progression. Without that record, the role risks losing continuity. With it, each new leader begins with a foundation that allows them to build higher and stronger.

Cultivating this future requires more than documentation. It requires building a pipeline of talent. Deputy CDAOs, field CDAOs, and domain leaders are not merely support staff. They are leaders in training who can test new approaches and inject fresh perspectives. As highlighted in the chapter on organizational design, distributed leadership is essential for resilience. By entrusting emerging leaders with meaningful responsibility, current CDAOs ensure that the institution remains strong even as individual tenures end.

Inspiring the next generation also requires broadening our sense of who can lead. Some CDAOs will come from technical fields. Others may emerge from operational roles, policy domains, or communications backgrounds where they first learned how data shapes outcomes. What matters is not a single career path but the ability to integrate across disciplines and lead with both technical literacy and organizational fluency. A diversity of perspectives will strengthen the institution and equip it to address challenges that no one discipline can solve alone.

There is also a moral responsibility. The leaders who follow will inherit tools of greater power, deeper integration, and wider consequence than any we know today. They will be asked to manage systems that make decisions at scale, to guard against risks that may not yet be fully understood, and to ensure that progress does not come at the expense of trust. Preparing them for this responsibility requires honesty and humility. By sharing what worked and what failed, today's CDAOs equip the next generation to lead with both courage and wisdom.

Finally, inspiring the next generation requires imagination. Those who step into this role tomorrow will not simply inherit the structures of today. They will adapt and extend them, applying concepts such as everyday modernization, adaptive governance, and mission agility in contexts we cannot yet predict. The responsibility of today's CDAOs is to create the conditions where imagination can flourish; lowering barriers, providing access to resources, and offering a vision of leadership that is both practical and profoundly meaningful.

REFLECTION
The Journey Through This Book

This book has traced the full arc of CDAO leadership. We began with the recognition that data and AI are no longer peripheral concerns but central to mission and enterprise success. We explored the foundations of strategy and policy alignment, the discipline of governance, the power of modernization, and the necessity of organizational design. We examined ecosystems that fuel innovation, playbooks that drive deployment, metrics that tie performance to outcomes, and ethics that ground every decision in responsibility. Each chapter built upon the last, creating a comprehensive framework for understanding what the CDAO must know, do, and become.

The threads are now woven together. The picture is clear. The CDAO is not a passing experiment. It is an institution in formation, a leadership role shaping the next generation of organizations, and a movement that will define how humanity uses its most powerful tools.

A Call to Action

We stand at the edge of an era that will be remembered for how leaders chose to use data and artificial intelligence. It will not be remembered for the technology alone. It will be remembered for the choices we made about how to apply it, govern it, and build trust around it. It will be remembered for the courage of those who chose to lead when the path was uncertain and the stakes were high. That is the essence of the Chief Data and AI Officer.

The call to action is simple: step into the role with conviction. Whether you are a current CDAO, an aspiring one, or a leader who will work alongside them, the moment requires you to act with clarity and courage. The future will not wait. The forces of technology, organizational change, and societal expectation are already converging. The question is not whether they will shape the future, but whether you will help guide that shaping.

To embrace this role is to accept responsibility for building trust where skepticism remains, for creating clarity where confusion reigns, and for fostering agility where rigidity once held sway. It is to recognize that the work you do will ripple outward, affecting not only your organization but the larger ecosystem of practice. It is important to understand that the standards you set today will become the norms that govern tomorrow.

And it is to embrace possibility. This is an extraordinary opportunity. Data and AI are not simply tools of efficiency; they are engines of discovery, catalysts for innovation, and instruments of progress. They have the power to transform how we defend nations, care for the sick, educate future generations, steward resources, and connect communities across every boundary that once divided us. They can extend human capability in ways once unimaginable. But only if we lead with purpose. Only if we treat this moment not as a technical challenge to be solved, but as a leadership challenge to be embraced.

The final message of this book is one of empowerment. You do not need to have every answer. You do not need to eliminate every risk before you begin. What you need is the conviction to start, the humility to learn, and the resilience to adapt when reality proves different from your plans. The CDAO role is still young, and that is its strength. It is not weighed down by decades of precedent that constrain imagination. It is alive with possibility. You have the chance to shape it, to define it, and to ensure that it becomes a force for good that will outlast your tenure.

So let this be your charge: take what you have read here, adapt it to your context, and make it real. Do not let these pages close without deciding on your next step. Start a conversation with your peers about what AI-ready means for your organization. Reexamine your policies and structures through the lens of adaptive governance. Launch a pilot that tests the principles of everyday modernization. Build a partnership that expands your innovation ecosystem. Mentor someone who could one day step into this role. Each step, no matter how small, moves you closer to the AI-ready enterprise and to the future that this role was created to build.

The world is watching. The pace of change will only accelerate. And the responsibility of leadership has never been greater. But so too has the opportunity. To be a CDAO, or to work alongside one, is to hold in your hands the ability to shape not only organizational outcomes but the trajectory of how humanity will use its most powerful tools. That is a responsibility worthy of your best energy, your sharpest thinking, and your most profound commitment.

The technologies we have discussed—data platforms, AI systems, analytics engines, and automation tools—are not the story. They are instruments. The story is what we choose to do with them. The story is written by leaders who refuse to accept that the old ways are the only ways, who see complexity as an invitation rather than an obstacle, and who understand that the organizations they serve deserve better than incremental improvement when transformation is within reach.

The CDAO role exists because the world changed faster than our institutions. It exists because data became too valuable to leave scattered, because AI became too powerful to deploy without wisdom, because the gap between what technology could do and what organizations actually achieved grew too wide to ignore. Into that gap stepped a new kind of leader (part strategist, part technologist, part diplomat, part visionary) armed with the conviction that organizations can be smarter, faster, more responsive, and more trustworthy if someone is willing to do the hard work of making it so.

That leader is you.

The path will not be easy. There will be skeptics who doubt the data's trustworthiness. There will be obstacles: legacy systems that resist modernization, cultures that fear change, and budgets that never seem adequate to the ambition. There will be moments when progress feels impossibly slow, when pilots fail, when the politics of organizations threaten to overwhelm the promise of technology. Every CDAO who has walked this path knows these challenges intimately.

But they also know something else. They know the moment when a dashboard finally tells the truth about operations, and leadership can act on it. They know the satisfaction of watching an AI model identify a pattern that humans missed, saving time, money, or lives. They know what it feels like to build a team that believes in what data and AI can accomplish, to see junior staff grow into leaders, to hand off a capability that will outlast their own involvement. They know that the work matters.

This is the work you are called to do.

The future of the CDAO is not something that will happen to you. It is something you will create. Every governance framework you establish, every pilot you launch, every conversation you have about ethics and trust and mission value—these are the bricks with which the institution is built. The CDAO of ten years from now will inherit what you create today. The organizations that succeed in the decades ahead will do so because someone chose to lead right now.

Choose to lead.

Step into the role with courage. Bring others along with you. Build the structures that will outlast your tenure. Develop the people who will carry the mission forward when you are gone. Tell the story of what data and AI can accomplish in ways that inspire rather than intimidate, that invite participation rather than demand compliance, that make the abstract concrete and the possible feel inevitable.

The journey does not end here.

It begins here, with you.

www.ingramcontent.com/pod-product-compliance
Lightning Source LLC
La Vergne TN
LVHW051056210926
848486LV00002B/122
9781960907127